# THE SURVIVAL
# FACTOR

# THE SURVIVAL
# FACTOR

## MIKE AND TIM BIRKHEAD

## Facts On File
*New York • Oxford • Sydney*

# Acknowledgements

The majority of the photographs appearing in this book are reproduced with the kind permission of the Survival Anglia Picture Library, as follows: Andrew Anderson pages 167, 173, 177, 182; Terry Andrewartha page 164; Jen & Des Bartlett pages 2, 81, 83, 95, 102; Joel Bennett page 116, 189; Joe B. Blossom/WFT page 144; Liz & Tony Bomford page 184; Ceballos/Kemps page 198; Ashish Chandola page 69; Bruce Davidson pages 87, 180, 205; Jeff Foott pages 10, 15, 21, 47, 60, 63, 73, 88, 90, 97, 104, 105, 107, 108, 113, 115, 119, 121, 123, 125, 127, 147, 148, 150, 151, 152, 153, 155, 156, 157, 160, 161, 166, 179, 181, 183, 190, 193, 197; Frances Furlong pages 134, 136; Dennis Green pages 18, 54, 67; Peter Hawkey page 141; M. Kavanagh page 52; Richard and Julia Kemp pages 158, 200; Chris Knights page 42; Dr F. Köster page 27; Mike Linley pages 72, 79, 96, 100; Lee Lyon page 86; Jozef Mihok page 175; Dieter & Mary Plage pages 37, 39, 41, 44, 46, 65, 93, 95, 192; Annie Price pages 129, 131, 132, 137, 163; Mike Price pages 34, 38, 59, 76, 91, 203; Rick Price page 146; Alan Root pages 103, 195, 196; Michel Strobino pages 62, 112, 124, 171; Maurice Tibbles pages 16, 55, 61; Mike Tracey page 159; Joanna Van Gruisen page 110; Ian Wyllie pages 22, 25, 26, 29, 32. Kind thanks are also due to Sue Harrison and Kathryn Shreeve of the Survival Anglia Picture Library for their help in making this picture selection.

The publishers would also like to thank the following for kind permission to reproduce their photographs in this book: Mike Birkhead pages 6, 9, 10, 11, 12, 13, 14, 31, 49, 50, 56, 64, 101, 126, 133, 138, 139, 140, 169, 170, 186, 201, 202. Oxford Scientific Films/M. P. L. Fogden page 77; Oxford Scientific Films/Roger Jackman page 85.

The authors would like to thank Cheryl Brown and Jeremy Bradshaw for their help in producing this book; David Quinn for his superb line drawings; and Sarah Mahaffey, Cindy Buxton, Peter Jennings, Mike Linley, and Malcolm Penny, Ken Smith, Richard Tinsley, and Ian Wyllie for their constructive comments on parts of the text.

The illustrator would like to thank Glen Fisher, Dr Malcolm Largen and Dr Ian Willis of The Zoology Department at National Museums and Galleries on Merseyside for their kind assistance.

First published in Great Britain in 1989 by Boxtree Limited

First published in the United States of America in 1990 by Facts On File Inc

Facts On File, Inc.
460 Park Avenue South
New York NY 10016
USA

Designed by Groom & Pickerill
Line drawings by David Quinn
Typeset by Cambrian Typesetters
Origination by Culver Graphics Litho Limited
Printed in Spain by CAYFOSA Industria Grafica, Barcelona

Facts On File books are available at special discounts when purchased in bulk quantities for businesses, associations, institutions or sales promotions. Please contact the Special Sales Department of our New York office at 212/683–2244 (dial 800/322–8755 except in NY, AK, or HI).

**Library of Congress Cataloging-in-Publication Data**
Birkhead, Mike.
  The survival factor.
  1. Animal defenses. 2. Adaptation (Biology)
3. Animal behavior. I. Birkhead, T.R. II. Title.
QL759.B57  1990  596'.05  89–26062
ISBN 0–8160–2355–7 (alk. paper)

# CONTENTS

# INTRODUCTION

A scene from one of Survival Anglia's most memorable wildlife documentaries shows two powerful male sea lions fighting on a beach in Patagonia, Argentina. The battle is for exclusive sexual access to a harem of females and is savage and relentless. Although evenly matched, one of the sea lions eventually gets the upper hand, pushing his rival, with ferocious bites, back towards the sea. Suddenly an enormous killer whale lunges from the surf and grabs the retreating animal. Within seconds, this sea lion is killed by the killer whale – one of the most powerful predators of the ocean.

The remaining sea lion is both survivor and winner. It has overcome its rival, escaped the clutches of a killer whale and earned the right to mate with the harem of females. Not only has it survived but it now has the opportunity to reproduce, to pass on its genes to the next generation. From an evolutionary perspective, this, for all animals, is their only purpose in life.

In 1859 Charles Darwin published *The Origin of Species by means of Natural Selection* – his theory on the workings of evolution. The subject of *The Survival Factor* series, and of this book, is Darwinian evolution and the way in which it affects all living organisms. Central to this discussion is the process known as natural selection, which shapes virtually every aspect of an animal's life – from the size of its feet to its mating behaviour. The special characteristics and adaptations that enable an animal to survive and reproduce are its 'survival factors'.

The story of evolution – the process of change by which reptiles developed into birds and by which a chimpanzee-like primate may have paved the way for the appearance of human beings – has been described in innumerable books. We are not concerned here with these major

*Opposite The male baboon's large canine teeth are used to attack prey animals like small gazelles; as a defence against predators; and to threaten its rivals.*

7

*A killer whale lunges forth from the waves and snatches up a sea lion from the beach.*

evolutionary steps but with the 'fine tuning' processes of nature, the way creatures are adapted to their environment.

Evolution does not necessarily take millions of years. Change occurs as the result of sexual reproduction, and every new creature that is produced is genetically unique. Variations between one individual and another are usually of little consequence, but now and then an unexpected divergence will appear. Take the mosquito, for example. These noxious insects have been sprayed with pesticides such as DDT for the past thirty years or so. Yet not all of them have disappeared. Initially, all but a very tiny number died instantly. However, not all mosquitoes were genetically identical and those few that survived had a slightly different make-up – they were immune to the effects of the pesticides. The unusual trait or 'survival factor' that kept these mosquitoes alive was subsequently passed on to their offspring which multiplied.

*This long-necked antelope – the gerenuk – did not get its elongated neck by constantly stretching up to reach the tender shoots and leaves at the tops of the bushes. It is the result of natural selection.*

Transformations such as these are often brought about by a natural change in the environment. This may be gradual and imperceptible, or dramatic, as after a volcanic eruption or the flooding of a river. However, no matter how well adapted an animal has become through natural selection, human interference can threaten to obliterate them in a matter of years if they cannot somehow avoid it. The massive African black rhinoceros is an example. Until recently this animal seemed to possess all the necessary survival factors to flourish in its natural habitat. The animal was enormously strong, and although its eyesight was poor, it had remarkable powers of smell and hearing, and its horn was a very effective means of defence. However, it has been increasingly hunted for that horn which is now commercially valuable. In 1970 the black rhino population stood at more than 65,000. Today it is down to 4,000.

*Though well adapted to their environment black rhinos are a threatened species today due to poaching by man.*

The process of natural selection emphasises those attributes which aid survival. Such factors can include the spectacular plumage of the vulturine guinea fowl, the camouflage colours of the bustard or the strange co-operative relationship between the oxpecker and the warthog. Each chapter of this book describes and illustrates one animal or group of animals, and examines the special features that help them to survive. Thus the hooves of the mountain goat which adapt to fit any rock shape are ideal for clambering along cliff-faces, and the probing tongue of the woodpecker is tailor-made for extracting insect food from rotting bark. These and many other survival factors have been 'picked out' through the rigorous process of natural selection.

Nevertheless, natural selection or 'survival of the fittest' is only part of the story. Darwin's observations led him to the conclusion that some animals reproduced more successfully than others. He believed that such differences were due to characteristics that gave certain individuals an

*The exotic plumage of the vulturine guinea fowl of the arid regions of East Africa is (unusually for birds) shared by both males and females.*

11

advantage over others of the same species in obtaining mates – larger antlers in a stag or a more showy tail in a peacock. If such attributes enabled a male to gain the favours of a female, they were likely to be perpetuated in the species. It worked in two ways, most clearly where males were competing for females, and less obviously where females were choosing particular males to mate with. Darwin called this phenomenon 'sexual selection'.

*This male black-bellied bustard's plumage provides camouflage when required but its colours also impress females.*

Peacocks are not the only species in which the male has a longer and more colourful tail than the female. This difference between the sexes also occurs among widow birds, tropic birds and the European swallow. The simple fact that some male swallows have longer tails than others may enable females to pick out the best males; it is not known for sure, but it is thought that tail length in some way gives an indication of the quality of the male to the female. Experiments carried out by zoologists have shown that these males do procure mates more easily than rivals with shorter tails, and also produce more young. Furthermore, female swallows, which often mate with more than one partner, continue to show preference for males with long tails. The outermost tail feathers of male swallows are longer than those of the females. To discover why this might have evolved in males, a Danish zoologist, Anders Møller, glued additional bits of tail on to certain males, whilst shortening the tails of others. Møller discovered that the males with the longest tails, artificial or otherwise, still enjoyed the greatest breeding success, and that the females seemed irresistibly drawn to them. The longer tail, therefore, has arisen through sexual selection – an adaptation that accommodates the female's preference.

*These ground forest hogs attract red-billed oxpeckers. Both benefit from this association for these birds feed on the hogs' tick-infested hides. The oxpeckers also warn of any approaching danger by uttering loud warning cries.*

*The long tails of these spectacular white-tailed tropic birds of the Seychelles attract females.*

*Opposite Male lions need large canine teeth not only to grip and kill their prey but also to fight rival males for the 'right' to a pride of breeding females. More than one male normally consorts with a pride of lionesses, but they are usually brothers.*

Many of the survival factor examples discussed in this book are more straightforward than the story of tail length in swallows. The special shape of an ape's or a monkey's hands enables it not only to grip objects, but to manipulate them too. Perhaps this is one reason why we see the use of tools in such creatures. It is relatively easy for us to appreciate the importance of delicate hands and manipulative digits when we ourselves, or our children, first try to write with a pen or make a model. But it is important to remember that the purpose of nearly every adaptation or characteristic which has evolved through natural selection is to increase an animal's potential to survive and to breed. A baboon's enormous canine teeth, for example, are used to defend themselves against predators like leopards, to kill prey like young gazelles, and to threaten and fight other baboons. If the teeth are too long they may hinder the animal whilst it is feeding – if they are too short they may not be good enough weapons. Only those baboons which have evolved the teeth of the 'right' length will be at an advantage, and will out-survive those that don't. The same principle applies to the lion's canine teeth, the diving ability of the seal, the parasitical habits of the cuckoo and the hunting powers of the eagle. They are all survival factors.

# The
# Cuckoo

The habits of the cuckoo are well-known. Ogden Nash, the 20th century American poet, probably summed up popular attitudes about this bird when he wrote:

> *Cuckoos lead bohemian lives,*
> *They fail as husbands and as wives –*
> *And therefore they cynically disparage*
> *Everybody else's marriage.*

But as we shall see they are most successful at breeding, sometimes producing as many as 25 offspring in one season.

The cuckoo is what is known as a brood parasite – it lays its egg in the nest of a different, host, species of bird and then has nothing further to do with it. The entire burden of hatching, feeding and caring for the young cuckoo falls on the hapless foster parents. Not only that, but on hatching, the young cuckoo throws out the host species' own eggs or young so that it has no competition for the foster parents' time and attention.

Not all cuckoos are brood parasites and not all brood parasites are cuckoos. About 80 species of birds worldwide have dispensed with the bothersome business of raising their own young and only about 50 of these are cuckoos of one sort or another. The remainder include the African whydahs and indigo birds, the cowbirds of North and South America, the honeyguides and even a duck – the black-headed duck of South America. Conversely, there are many species of the cuckoo family which build a nest and incubate their own eggs in the conventional fashion.

All brood parasites show a range of fascinating adaptations in support of

Opposite *A young cuckoo being fed by its foster parent, a tiny wren. A cuckoo chick will have outgrown the host bird within days of hatching.*

17

their unorthodox life style, but the best-known and most fully studied is undoubtedly the European cuckoo which is the species we shall concentrate on in this chapter.

The European cuckoo's strange behaviour was noted as long ago as 300 BC by the Greek philosopher Aristotle, who also recorded the young cuckoo's ejection of the host bird's eggs from the nest. It is unlikely that many Europeans would have had access to Aristotle's writings, but the cuckoo's habits were certainly well enough known during the Middle Ages for them to be mentioned by Chaucer (in *The Parlement of Foules*, 1382), and for the term 'cuckold' – describing a man deceived by his wife –to have passed into the English language. However, it was not until scientific ornithology began to get into its stride in the 19th century that some of the more remarkable aspects of the cuckoo's behaviour began to be widely known. They were considered so remarkable that many people, biologists among them, were unable to believe that they arose through natural selection. The cuckoo's adaptations were simply too perfect to have evolved bit by bit, in piecemeal fashion. It was thought they must have been conferred all at once in an act of divine creation. However, such naive views, as we shall see later, were to be proved incorrect.

Early in the 20th century, significant contributions were made to our knowledge of the cuckoo. These were mainly through the work of one man, Edgar Chance. Chance was a businessman rather than a biologist, but in his spare time he collected birds' eggs, particularly those of the

*The European cuckoo, with its grey plumage and barred underside, bears a striking resemblance to the predatory sparrowhawk, pictured below.*

cuckoo and its hosts. During the 1920s and '30s, Chance had the now dubious honour of being one of the best-known egg-collectors in Britain. His detailed observations helped to dispel popular misconceptions about the habits and behaviour of the cuckoo. He himself would probably have claimed that his greatest achievement was to be able to predict accurately where and when a particular cuckoo would lay its eggs. It was this that made it possible for the whole of the cuckoo's breeding cycle to be filmed, resulting in an early classic of wildlife photography, *The Cuckoo's Secret*. Since Chance's day, other researchers have continued the work. Most significant in recent years have been the field experiments carried out by the British ornithologists Nick Davies and Mike Brooke from the University of Cambridge, and the field observations of another British ornithologist, Ian Wyllie. As a result, we now have a very detailed picture of the lifestyle of the cuckoo and can recognise it for what it is: not a case of divine creation, but a superb example of the complex adaptations that occur in the intricate interrelationship between a brood parasite and its host.

*The cuckoo's zygodactyl toes, with two pointing forwards and two pointing back. Species such as blackbirds and sparrows have three pointing forwards and one pointing back.*

The cuckoo is a medium-sized bird, weighing about 120 g (4.2 oz) – slightly smaller than a pigeon. In common with toucans, parrots and woodpeckers, cuckoos have two toes pointing forwards and two pointing back. This is thought to facilitate moving around on twigs and branches which, for a bird that lays its eggs in nests that may be hidden in thick vegetation, is an advantage. Male and female are very similar in appearance; the male is slightly larger, but the difference is usually imperceptible. A small proportion of females have a rusty-red plumage, distinct from the normal grey colour. There is no male equivalent, and nor does there appear to be any particular explanation for the colour variation.

With its grey plumage, barred underside and hawk-like shape and silhouette, the European cuckoo looks deceptively like the predatory sparrowhawk – the two are often confused. This resemblance may not be a coincidence. The cuckoo cannot lay its egg while the future foster-parent is in residence; it has to pick a time when the parent bird has left the nest – and the eggs – so that it can perform its surreptitious substitution. The appearance of a hawk-like bird may provide just the encouragement the potential hosts need to make them abandon the nest for a while and take cover elsewhere. This idea is supported by the fact that several species of cuckoos in other countries appear to mimic the appearance of hawks that are common locally.

Another possibility is that, by looking like a hawk, cuckoos are protecting themselves from predation – a hawk is less likely to attack another hawk than some other, easier prey. The cuckoo, especially the male when singing, spends a great deal of time sitting on exposed perches, making it particularly vulnerable. Not only does it need to keep a constant lookout for the large hairy caterpillars that it prefers to eat (particularly those of the magpie moth which other birds find distasteful), but it must also keep a close eye on the activities of potential hosts. Mimicking a hawk could also be useful in Africa, the European cuckoo's wintering ground, where birds of prey are numerous.

*Hairy caterpillars, such as this 'woolly bear' or garden tiger moth caterpillar, are usually distasteful to birds, but not to the cuckoo.*

The European cuckoo's breeding range extends across most of Europe and Asia. It winters in Africa south of the Sahara and arrives in Britain around mid-April. The cuckoo's year is well described by a traditional rhyme:

> *April come I will.*
> *In May I've come to stay.*
> *In June I change my tune,*
> *And in July, away I fly.*

Observation shows that this is broadly correct. Certainly, cuckoos spend a very short time in Britain (12–15 weeks) – less than any other migrant. Despite this, it is on their breeding grounds that they have been most studied. Little is known about cuckoos in their African winter quarters.

The mellifluous and mysteriously ventriloquial call of the cuckoo is a sign that spring has arrived. The call is uttered only by the male cuckoo,

who arrives slightly before the female, and begins to call as soon as he reaches the breeding grounds. The 'song' consists of about ten 'cuck-oos' separated from each other by one- or two-second intervals. Some individuals carry on for longer, and one has been heard to call 270 times without a break. The 'song' is delivered from a prominent perch and calling is most frequent early in the day. Its main function is to attract a female: male cuckoos do not appear to be territorial since the areas over which males sing often overlap as do the egg-laying ranges of females.

Cuckoos do have other calls. Both sexes give a 'wah-wah' cry, especially during aggressive encounters with other cuckoos. The female also has a 'bubble' call consisting of about 15 notes on a descending scale. This call is most often given just after the female has laid an egg and may be used to attract males for further mating.

Although generally not territorial, cuckoos return regularly to the area in which they bred in previous years. One of Edgar Chance's cuckoos, recognisable by its distinctive eggs, returned for no less than five consecutive years to the same common in southern England.

As the ranges of the males overlap, it is hardly surprising that the females are regularly courted by several males during the course of the breeding season – a female may have as many as four males in pursuit at the same time. The male courts the female by singing and posturing, often with a small twig or leaf in his bill. Mating itself has very rarely been observed. Ian Wyllie, who studied cuckoos extensively in Cambridgeshire reed-beds, saw it just three times over a period of six years after thousands of hours of

*In addition to certain species of cuckoo, there are about 30 species of birds worldwide that are also brood parasites. Above, the North American yellow warbler plays host to a brown-headed cowbird chick, as well as raising her own hatchlings.*

21

observation. Both male and female call before mating, the male with his 'cuck-oos' and the female with her 'bubble' call. The male approaches the female in a gliding flight, still calling as he does so. He lands directly on the female's back, and mating takes place.

It is not known for certain if the male and female form a pair bond. Ian Wyllie considered that they were promiscuous, with both male and female mating more or less indiscriminately with several partners. This seems likely. Although most birds are monogamous, promiscuity is not unusual among brood parasites.

After mating, the female is ready to lay her egg. This act of stealth and deceit is far from straightforward, and it is now that we begin to see some of the more subtle and complex adaptations of the cuckoo to its parasitic way of life.

□

The first problem that faces the cuckoo is to find a nest belonging to the right species of host. Not all birds fall victim to the cuckoo – many small birds that might appear to be ideal cuckoo hosts will not tolerate any strange egg in their nest and will remove it straight away. Thus, while a hundred or so European species have been recorded as hosts, only a handful are used regularly. In Britain the main hosts are the reed warbler, meadow pipit, dunnock and pied wagtail. However, for each female, the choice is restricted further by the phenomenon known as egg-mimicry.

The cuckoo's ability to produce eggs that bear a striking resemblance to those of its host is one of the bird's best known adaptations. It was once thought that the cuckoo could change the colour and pattern of its eggs at

*A cuckoo's egg is conspicuously larger than the host's eggs to our eyes but the egg is sufficiently similar in pattern and colour not to be detected by the foster parents.*

## PARASITISING YOUR OWN

The success of a brood parasite like the European cuckoo depends on how well it can 'dupe' its host, whether reed warbler or dunnock. This often involves a whole range of elaborate deceptions. How much simpler it would be to parasitise a member of your own species.

This is exactly what a number of species of birds have done – golden eye, European starling, moorhen and North American cliff swallow among them. Until recently, it was not appreciated how widespread this kind of behaviour is, mainly because of the difficulty of egg-detection. It is one thing to distinguish a cuckoo's egg from those of its host, but quite another to tell apart the eggs of two different starlings.

Most of the birds that parasitise their own

species lay one or more eggs in the nests of other birds, and then lay and rear a separate clutch themselves. They are not always very successful. Moorhens, for example, are rather poorly adapted to parasitising their own. They dump their eggs in other nests at any stage of incubation, with the result that the success rate is low. Few extra offspring are produced.

The cliff swallow, however, is much more sophisticated. In one respect it is unique: it has developed the ability to move eggs from its own nest to another by carrying them in its beak. It also times the transfer of its eggs extremely carefully so that most of them hatch in the hosts' nests at exactly the right time. As a result, this can be a very successful strategy for some swallows, allowing them to produce extra offspring at very little cost to themselves.

*A recent discovery has shown that some North American cliff swallows actually carry eggs from their own nest to other swallows' nests nearby. They will then return to their own nest to lay and rear a clutch. Such a strategy increases the reproductive success of these individuals.*

will, allowing it to lay in a range of different nests. This is now known to be impossible. The female cuckoo can mimic only one type of egg and must try to lay in a nest of this species if her offspring is to have a good chance of survival. The degree of colour mimicry is often startling, and has resulted in the cuckoo showing perhaps the greatest degree of colour variation in its eggs in the entire avian world.

There are, therefore, 'reed warbler' cuckoos, 'meadow pipit' cuckoos, 'pied wagtail' cuckoos and so on. The major, and as yet unexplained, exceptions are the 'dunnock' cuckoos, whose eggs do not resemble those of their host at all. At one time it was thought that this was because the dunnock was a recent cuckoo host, but Chaucer's reference in *The Parlement of Foules* makes it clear that the dunnock was a popular cuckoo host as long ago as the 14th century. Whatever the case, the dunnock seems tolerant of colour variation in its eggs, so perhaps the cuckoo has no particular need to mimic them.

Just how important the selection of the right host can be for the cuckoo has been demonstrated by Nick Davies and Mike Brooke in a series of ingenious experiments. They made realistic models of different types of cuckoo eggs (realistic enough to fool a leading British ornithologist who unwittingly recorded one as part of a clutch he discovered) and placed them in the nests of reed warblers to examine the response of the hosts. When model 'pied wagtail' or 'redstart' cuckoo eggs were placed in the nests, almost all were ejected. When model 'reed warbler' cuckoo eggs were used, none of them was thrown out. It would seem that egg mimicry in cuckoos has evolved to its present level of sophistication because poorly matching eggs are ejected by the host.

The female cuckoo is thus very restricted as to where she can lay her eggs, and it is important that she keeps an eye on the whereabouts of potential hosts. Furthermore, since she normally lays around 12 eggs in a season at a rate of one egg every two days, she needs to know the locations of several nests simultaneously. She also needs to know what stage the host birds have reached in their own nesting cycle because the timing of her laying can be crucial. Altogether, the female cuckoo has a great deal of spying to do.

Timing can be almost as important as selection of the right host species. Almost all cuckoo hosts, like the reed warbler, lay one egg each day until their clutch of around five eggs is complete. The best time for the cuckoo to lay its egg is after the hosts have started laying and before they have finished. Ian Wyllie found that, of 90 reed warbler nests that had cuckoo eggs in them, 86 (95 per cent) were parasitised during egg-laying and only four after the clutch was complete. The dangers of laying too early or too late are simple. If the egg is laid too early, as Davies and Brooke found with their model cuckoo eggs, the host throws it out or deserts the nest. If the egg is laid after the host has begun incubation, the young cuckoo will probably hatch too late to eject its nest mates. A cuckoo that finds a host's nest where incubation has already started will often rob the nest causing the unfortunate host to start a new clutch, thus providing the cuckoo with a further chance for parasitism.

Careful timing by the cuckoo is important in another respect. One of Edgar Chance's main findings was that, unlike most other birds which lay in the morning, cuckoos lay their eggs in the afternoon. This gives the cuckoo more time to find and watch a suitable nest. If she laid at dawn, like most birds, she would have to have prepared the day before. Davies and Brooke found that model eggs that were placed in the host's nest in the morning were more likely to be ejected than those placed there in the afternoon.

Even when she has selected a place and a time, the female cuckoo has still got the problem of introducing her egg into the host's nest in such a way that it won't be rejected. At this stage, yet another set of variations in the cuckoo's virtuoso repertoire of adaptations comes into play.

The cuckoo prefers to lay its egg while the hosts are away from the nest. If a laying cuckoo is seen by the hosts it is vigorously mobbed and chased and, more importantly from the cuckoo's point of view, its egg is more likely to be ejected. However, Edgar Chance showed that a cuckoo using a meadow pipit's nest is nearly always attacked while she is laying. Generally though, prior to laying an egg, the female cuckoo will watch the intended host's nest for several hours, sitting quietly in a tree or bush, waiting for her opportunity. During this time the egg passes along her oviduct (egg-laying tube) so that it is ready for immediate extrusion when she arrives at the nest. When the host leaves the nest, the cuckoo makes its approach in a long, silent hawk-like glide. If, however, the host appears reluctant to leave the nest, the cuckoo has a more direct approach.

We saw this ourselves during the filming of a television sequence on the laying behaviour of the cuckoo. A female cuckoo, ready to lay her egg, had apparently fixed on a particular reed warbler's nest as her destination. Unfortunately for her, the host bird showed no inclination to leave the nest. So, instead of approaching silently, the cuckoo crashed noisily into some nearby vegetation, causing a considerable commotion and frightening the host bird away.

*A dunnock plays host to a young cuckoo. Though cuckoos are well known for laying their eggs in the nests of other species, more species of the cuckoo family build nests and incubate their eggs than do not. There are in fact 80 species that act as true parents to their young.*

What followed was, in its own way, even more remarkable. The cuckoo grabbed the side of the nest, picked up the reed warbler's egg in its beak and flew off. The whole sequence of events lasted only eight seconds, and yet in that time the cuckoo had also managed to lay its own egg. Most birds take about 20 minutes to lay an egg. The ability of the cuckoo to lay its egg so quickly is a major adaptation of brood parasites.

For their body size, cuckoos lay remarkably small eggs. The average cuckoo's egg weighs just 3 g ($\frac{1}{9}$ oz), while most birds of the same size lay eggs weighing about 10 g ($\frac{1}{3}$ oz). This must be a factor in the cuckoo's ability to lay its eggs so rapidly, but it also has other advantages. One is that small eggs hatch sooner than large ones. It is vital for the young cuckoo to hatch either before, or at the same time as, its nest mates. The combination of careful timing by the female cuckoo, together with a short incubation

*Swooping onto a reed warbler's nest, this female cuckoo removes one of the host bird's eggs and lays one of her own to replace it in a matter of seconds.*

period, ensures that on most occasions this happens. The incubation period of a cuckoo's egg is about 11 days, while that for host species is slightly longer – 12 days for the reed warbler and dunnock, 13 to 14 days for the meadow pipit, pied wagtail and redstart.

Small eggs also mimic those of the host species much more closely. When Nick Davies and Mike Brooke placed 10 g-sized model 'reed warbler' cuckoo's eggs in reed warbler nests, they found that about half of them were ejected, whereas none of the model 3 g normal-sized ones were. Finally, laying such small eggs may enable the cuckoo to produce more of them. Although around 12 eggs in a season is normal, one female observed during filming of a television programme laid 25.

There is one other bizarre adaptation used by the female cuckoo in laying her eggs. The opening of the oviduct can be protruded from the body, allowing the bird to 'squirt' her egg into position. The eggs may land with a bit of a thump, but there is another adaptation to take care of that:

cuckoo eggs have unusually thick shells. The advantage of this method of delivery is that, as well as being quicker, it allows the cuckoo to use nests that would otherwise be inaccessible. Young cuckoos are sometimes reared in the most unlikely places. They have been found inside nest boxes with tiny entrance holes designed for tits, as well as in the delicate domed nests of such species as the wren. This is an advantage that the cuckoo is well advised to exploit with care – clearly a young cuckoo reared in a blue tit's nest box would never be able to escape when it was mature.

Before the cuckoo abandons its offspring to the care of the foster parents, there is one other action she usually carries out. This is a rather strange piece of behaviour, which gave rise in the 19th century to the belief that the female cuckoo swallowed her egg after laying it and then regurgitated it into the host's nest. The evidence for this came from birds

## ANIS

The anis are group-living cuckoos found in Central and South America. They are odd-looking birds with even odder habits. There are three species. The best known is the groove-billed ani, whose extraordinary breeding behaviour has been studied by Sandy Verhencamp in Costa Rica.

Groove-billed anis live in groups of up to four pairs. They share a territory as well as a communal nest. All the members of the group help to build the nest, and when it's complete all the females lay their eggs in this one 'basket'. Incubation is also shared. Such extreme sociability is unusual among birds; but the appearance of domestic harmony within the ani groups is deceptive. In reality, the females are locked in fierce and intense competition.

One thing that had puzzled early observers was the litter of ani eggs strewn beneath the nest. It was Verhencamp who discovered the truth – female anis throw each other's eggs out of the nest. Only a certain number of eggs can be incubated efficiently, and each female in the group tries to ensure that those eggs are hers. To achieve this, each female visits the nest before she starts laying and throws out one or more of the existing eggs. As soon as she starts laying herself, however, she has to stop doing this. A female ani cannot tell her own eggs from those of other females, so she would risk throwing out her own eggs if she did not.

There is, of course, great competition to be the last to lay, and the privilege is taken by the

*All three species of anis, including the smooth-billed ani, are group-living cuckoos. Several pairs share a territory as well as a communal nest. Above, an adult feeds the group's chicks.*

dominant female. This means that she can throw out the eggs of her nest mates, but they have little chance to throw out hers. The subordinate females do have one or two ways of retaliating – they tend to go on laying longer and produce more eggs than the dominant female, and they also produce the occasional late egg some time after they have laid the rest of their clutch. By adding to the clutch, they take the risk that some of the eggs won't hatch; but the more of their own eggs are in the nest, the better their chances of producing offspring. Nevertheless, the dominant female still ends up with most eggs.

# CHICK MIMICRY

The indigo birds and whydahs of Africa are brood parasites just like the European cuckoo, but there are some interesting differences in the ways in which they have adapted to a parasitic lifestyle. For one thing, they do not eject the eggs or young of their hosts – waxbills – from the nest. For another, instead of mimicking the eggs of their hosts, they mimic the appearance and behaviour of their young.

In both indigo birds and whydahs, the males are most brightly coloured than the females. Male indigo birds sport beautiful metallic blue plumage, while male whydahs sometimes possess spectacularly long tails.

Males have display or song posts to which they attract females for mating. A successful male may lure and mate with as many as 20 different females; less popular ones may fail to mate at all. As in many other polygamous birds, the male plays no part in nesting and neither, in the case of these birds, does the female.

After mating, the female indigo bird or whydah lays her egg in the nest of a waxbill. Each species parasitises a different species of waxbill: the village indigo bird parasitises the red-billed firefinch; the paradise whydah preys on the melba finch; and the straw-tailed whydah uses the purple grenadier. This link with a particular host is closely associated with chick mimicry. Young waxbills have complex gape and mouth markings which are mimicked exactly by the brood parasite concerned. Interestingly, there are three species of waxbill that share identical mouth markings and all are parasitised by the pintailed whydah.

Not only do the young whydahs mimic the appearance of the host's young, they also copy the way it feeds. Waxbills have a special feeding technique. The parent partially digests food in its crop. It then sticks its beak deep into the chick's mouth and literally pumps the food into its throat. In order to breathe during this operation, the young waxbill has to adopt a particular posture; young whydahs and indigo birds adopt this too.

A further refinement of chick mimicry is demonstrated by a tropical Asian cuckoo called the koel, which parasitises crows. Here the chicks mimic the plumage of the host's young.

The male adult koel also mimics the plumage of its hosts: the crows are vigorously territorial and while they are chasing away the intruding male, the female koel can sneak in and lay her egg. Koels also live in Australia but here they parasitise other species, and the young do not mimic the host. They do, however, eject their nest mates from the nest.

As well as looking like their host's young, the chicks of some brood parasites also sound like them. Young great spotted cuckoos reared by pied crows, and striped cuckoos reared by arrow-marked babblers, are both known to give begging calls identical to those of the host young. Vocal mimicry even occurs in a brood parasite that kills its nest mates. The young greater honeyguide, which parasitises red-throated bee-eaters, has a hooked tip to its bill with which it pecks its nest mates to death. Nevertheless, the young honeyguide still mimics the calls of the young bee-eaters.

*This shows the remarkable similarity that has evolved in the gape patterns of the chicks of two species of whydah birds and the chicks of the birds that they parasitise.*

Melba Finch

Paradise Widow Bird

Purple Grenadier

Straw-tailed Widow Bird

collected near the host's nest, whose throats were often found to contain an egg. The truth is that, as she is laying, the cuckoo removes and eats one of the host's eggs. It might be thought that this was an essential part of the cuckoo's deception – that by leaving the same number of eggs in the nest, the cuckoo might fool the host bird into thinking that nothing had happened. The evidence, however, seems to contradict this. Reed warblers are no more likely to eject a model cuckoo's egg when it is simply added to those already in the nest than they are when a substitution is carried out.

One possible explanation is that, if the cuckoo did not take an egg, the host might have too many eggs to incubate effectively. Certainly, reed warblers with an additional egg in the clutch often fail to hatch all their own eggs. But perhaps the most attractive explanation is that the cuckoo is simply snatching a quick meal. Generally, the cuckoo will not go on to take more than one egg, for, presumably, while the host tolerates the substitution or addition of an egg, it won't tolerate the loss of one. A missing egg is presumably an indication that a predator is around, and the bird deserts the nest.

With her egg laid, and with the bonus of one of the host's eggs to sustain her, the female cuckoo's maternal role is at an end. From now on, her offspring is on its own. Through incubation and fledging, it is entirely dependent on the care of its adopted parents. However, the mother does not leave the young cuckoo without instinctive resources, for the newly hatched chick demonstrates what is perhaps the most remarkable of all the numerous and varied adaptations of the cuckoo – the ruthless ejection of the host's eggs or nestlings from the nest.

*A cuckoo chick, less than one day old, instinctively completes the Herculean feat of ejecting one of its host's eggs.*

29

*Opposite A pair of reed warblers work flat out to keep the young cuckoo satisfied.*

The cuckoo's ejection behaviour was first described in detail by Sir Edward Jenner, the inventor of vaccination, in 1788. When his observations were first published, no one believed them. It is not hard to see why: even today the story seems incredible.

The young cuckoo hatches after about 11 days. Some eight hours later, and for the next four days, it sets about the task of ejecting its companions from the nest. Though still blind and naked, the chick manoeuvres itself inside the nest until an egg is lodged between its back and the side of the nest. Then, heaving with its legs and holding its tiny wings backwards to keep the egg in position, the young hatchling jerks the egg over the side of the nest. Chicks are treated in an identical fashion. The foster parents remain unconcerned as their own eggs or nestlings follow each other to the ground, even when they can see what is going on. During filming, we even saw eggs being pushed up and out of the nest while the reed warbler host was still sitting on them. The eggs simply plopped out in front of the host bird, who carried on incubating any remaining eggs as though nothing had happened.

Sometimes young cuckoos are too vigorous in their attempts to remove their nest mates and fall out of the nest themselves. Apart from this, their only problem seems to arise when two or, rarely, three female cuckoos choose to lay their eggs in the same nest. When this happens, the first hatched, or strongest, ejects the others. In one case, in which three different cuckoos parasitised the nest of a great reed warbler, one chick was thrown out soon after hatching: the other two then tried to eject each other for four days. Eventually, they both fell out. But in another case, described by Chance, a gardener found two cuckoo chicks in a dunnock's nest and both were reared successfully.

Very occasionally the young cuckoo fails to remove its host's eggs and the young cuckoo and chicks are reared together. This is most likely to happen where the host is a hole nester, like a robin, and where the eggs or chicks are difficult to eject.

If all goes well, however, the young cuckoo soon has the nest to itself, together with the undivided attention of its foster parents. It grows rapidly. By the time it is four days old, it is as heavy as one of its parents; after three weeks it weighs four times as much as they do, and dwarfs them both. By being reared alone, it gets all the food its foster parents can bring and it has a useful adaptation that ensures that the food keeps on coming – a huge, startlingly red gape. A nestling's gape, or wide open beak, provides a stimulus to the parents to feed it. The cuckoo's cavernous red gape may be irresistible to the overworked foster parents.

After about three weeks, the young cuckoo is ready to leave the nest. Even then, it is looked after for a further two or three weeks. Finally, it is ready for independence. It will take to the air and leave Britain for Africa where it will winter, before returning the next year to become a brood parasite in its own right.

The life cycle of the cuckoo raises the question of why host birds should accept a brood parasite's eggs, and why they should feed this alien monster in their nest? The answer is that they cannot help themselves. They are programmed by natural selection to behave in certain ways and cuckoos and other brood parasites are simply exploiting these fixed behavioural patterns. Natural selection favours those individuals that can tell when they have been parasitised and eject the cuckoo's eggs. This may already have happened in those species that will not tolerate a strange egg in the nest. On the other hand, the cuckoo is subject to natural selection too and will evolve strategies to defeat those adopted by its hosts. And the pressure of natural selection on the cuckoo is considerably greater than that on the host species. The cuckoo is dependent for its survival on its hosts. If all potential hosts were to reject it, the cuckoo would become extinct. The host species on the other hand could survive being parasitised since only a small proportion suffer the attentions of the cuckoo, and most of these will breed again to produce their own offspring. To put it another way, there is an evolutionary arms race between the cuckoo and its hosts, but the struggle is not an equal one. The cuckoo is always a few steps in the lead, because of its extraordinary survival factors. It is these adaptations which make it such a fascinating bird.

*Once fledged, young cuckoos stay around the nest site demanding more food from their foster parents for several days.*

# TOOL
# USERS

Technology is so pervasive, so much an intrinsic part of modern life, that we tend to take it for granted. It is easy to forget that it had to start somewhere. Yet the first crude tools our ancestors made – from flakes of rock, bits of timber and other natural materials around them – were the precursors of the computers, turbines, aeroplanes and all the other pieces of machinery essential to the modern world.

Until recently, it was considered that this ability to make and use tools was one of the defining characteristics of the human race. Tool using was considered an indicator of intelligence, and it was one of the things that separated *Homo sapiens* from the rest of the animal kingdom. As is so often the case, however, nature refuses to conform to such a convenient theory. In recent years, many examples of tool use in the animal world have come to light and there are no doubt others that are still waiting to be discovered. Animals are no strangers to 'technology' and some of the ways in which they use it are surprising as well as being remarkably sophisticated.

Before looking at some examples of tool using among animals, it would be as well to be clear about what we mean by the term. The concept is not as straightforward as it seems.

The nests of ants, bees, wasps and other social insects may appear to be marvellously complex examples of animal technology, but they are not examples of tool use. The most widely accepted definition of tool use is that it is the use of some external object as an extension of the body to attain an immediate objective. This definition would exclude a gull or crow that drops a shellfish from a height onto a hard surface. But it would include an Egyptian vulture that throws a stone at an ostrich's egg to break

*A baby orang utan gets to grips with a film camera! The ability to use tools is best documented amongst the primates.*

it open. The vulture is manipulating the stone and using it as an extension of its beak to crack the egg.

There are also different ways of interpreting tool-using behaviour. A bolas spider, swinging its sticky lure through the air to capture a flying moth, is behaving in a way that is innate. It hasn't learned this behaviour; it was born with it, though it gradually learns the finer points. Clearly, in this case, tool using is a part of the repertoire of adaptations that this particular species of spider has evolved. On the other hand, a thirsty chimpanzee in a Tanzanian forest may find some water in a tree hole, but be unable to reach it with its mouth. It will collect some leaves, chew them until they form a crude sponge, then soak the sponge in the pool so that it can squeeze the water into its mouth. In this case, the chimpanzee is showing what the psychologists would call 'insight'. It has identified a problem and has worked out a solution probably by trial and error; and the solution has involved the making and using of a tool. It has not inherited this behaviour; it has either worked it out for itself, or learned it by observing another chimpanzee.

Tool using thus covers a broad range of activities throughout the animal kingdom, from the mindless and mechanical to the considered and imaginative. It is also found at all levels, not just among our nearest evolutionary ancestors, the monkeys and the apes. In fact, examples of tool using can be found among even the most unpromising candidates – the invertebrates, those animals without backbones.

Ant-lions and worm-lions live in arid areas throughout the tropics. Although their names make them sound like large predators, they are actually tiny insect larvae. Ant-lions (*Myrmeleon*) are members of the lacewing flies, while worm-lions, which comprise two genera, *Vermilio* and *Lampromyia*, are true flies. Both have developed the same unusual method of trapping prey – they dig small pitfall traps for it.

The funnel-shaped pits are constructed in dry soil or sand using a novel excavation technique. The larva heaps sand or soil onto its head and jaws and then flicks it with some force up and behind it, working in a circle to produce a conical pit. Once the trap is complete the little animal burrows into the bottom of the pit and there lies in wait for unsuspecting prey.

Ant-lions, as their name implies, prey mainly on ants. The lion part of their names comes from the ferocity with which they attack any ant that comes within range of their huge jaws. Ants stumble into the pit by accident and, because it is constructed of friable soil or sand and has steep sides, they have to struggle to get out. The task is made more difficult, if not impossible, by the ant-lion. While the ant is scrabbling about trying to get up the sides of the pit, the ant-lion is busy chucking sand at it, again using its head and jaws as a 'catapult'. Eventually the shower of sand may knock the ant within the ant-lion's reach. Once seized, the ant is dragged into the sand at the bottom of the burrow where it is consumed.

The behaviour of worm-lions, which also flick sand at their prey, is essentially the same as that of the ant-lion. Worm-lions, however, also demonstrate another technical skill. When the larva is about to shed its skin and change into a pupa (which it does prior to emerging as a real fly), it exudes a sticky fluid. Grains of sand stick to this fluid which helps to hold the skin to the soil, thereby helping the pupating worm-lion to extricate itself. Without the sand particles to produce friction, the worm-lion would find it much more difficult to complete this crucial part of its development.

Several other insects attach extraneous objects or material to themselves, but for very different reasons. A particularly intricate example involves the larva of another lacewing fly, the green lacewing. The larvae of this fly feed on woolly alder aphids. Like many other aphids, the woolly alder has a

*An ant-lion larva lies in wait at the bottom of its specially constructed pit. It catapults grains of sand at passing ants in an attempt to knock them down into the pit and eat them.*

mutually beneficial arrangement with a particular species of ant – the ants protect the aphids from other insect predators and in return the aphids provide the ants with highly nutritious honeydew. So, if the green lacewing larva is to enjoy a decent meal of woolly alder aphids, it has to find a way of avoiding being attacked by the ants. It does this by disguising itself as an aphid, in order to avoid being detected by the ants. Woolly alder aphids are so named because they produce clumps of woolly wax from their bodies. The lacewing larva takes this from the aphids and sticks it on its own back, fooling the ants into thinking that it is simply another aphid. Experiments in which the wax is removed from the fly larvae show that the larvae immediately become vulnerable to attack from the ants, and they quickly try to re-cover themselves.

————————————— ☐ —————————————

As might be expected, some of the most intriguing examples of invertebrate tool use are to be found among those master-builders of the insect world, the ants, bees and wasps – collectively known as the Hymenoptera. Several species of ant of the genus *Aphaenogaster* use tools to carry soft, or semi-liquid foods. The tool consists of a piece of leaf, wood, sand or mud, with which the food is picked up. The ant then carries its 'plate' and food back to its nest. By using a tool the ant is able to transport ten times as much food per trip as it could have done otherwise.

The ferocious weaver ants of South-east Asia and South Africa use a highly unorthodox tool to help them construct their nests. The nest, located in the branches of a tree, consists of leaves stuck together – with silk. And the tool that they use to do this is their own larvae. The ants' larvae possess silk glands with which they spin their own pupal cocoons. The adult ants exploit this, using the larva like a tube of glue to stick the leaves of the nest together. Holding the larva in its jaws, an adult points the larva's silk glands at the leaves and, as the silk is extruded, the leaves, which are held together by many other ants, are glued. This brilliant piece of 'natural ingenuity' has evolved through natural selection. Although it may appear otherwise, such behaviour requires no conscious decision making by the ant.

The sandwasps *Ammophila* and *Sphex* also use a tool – a sort of miniature hammer – as an aid in construction. These wasps, which are common in sandy areas, dig a small burrow in which to lay their eggs. Each egg, or rather the larva that hatches from it, is provided with a supply of fresh food in the form of a caterpillar, which the female paralyses with her sting. After laying her egg and placing one or more caterpillars inside the nest, the wasp closes the burrow entrance with pebbles and soil. She then picks up a stone in her jaws and, vibrating her head at high speed, bangs down the soil. Sometimes the wasp may even insert a small twig into the soil and jiggle it about to help settle the material. The end result is that the burrow with its precious egg and living food supply is completely concealed.

A rather different form of technical skill is shown by a common and familiar resident of the seashore, the hermit crab. This uses the shells of whelks, winkles and other gastropods in much the same way as a medieval knight used a suit of armour – to protect the soft and vulnerable parts of the body from attack. Hermit crabs, however, do not simply put on a shell as if it were a new pullover; they show two remarkable adaptations to the problems of 'wearing' the shells of alien snails. First, unlike all other crustaceans, the hermit crab's abdomen is distinctly asymmetrical, coiling in a most bizarre fashion. Snails' shells are of course also asymmetrically coiled, so the hermit crab has plainly had to adapt its own anatomy in order to fit into its adopted shelter. Second, at the very tip of their abdomen they possess a hook-like structure which enables them to hold on to the shell. Anyone who has tried to remove a hermit crab from its shell will know how tenacious these creatures can be. Yet, when the time comes for a hermit crab to change shell, having outgrown its present one, it can accomplish the change over in just a few seconds.

Some hermit crabs, like *Dardanus*, seek yet further protection from more powerful predators and carry sea anemones, such as *Calliactis*, around on their shells. Sea anemones possess stinging cells in their tentacles, which

*Weaver ants use their larvae as portable tubes of glue, to stick together the leaves that form their nest.*

37

*A hermit crab carrying a sea anemone around on its shell. Exactly what the sea anemone gains from this relationship is not known, but crabs with anemones on their shells are known to be much less vulnerable to attack by octopuses.*

they use to protect themselves and to capture food. The crab effectively parasitises these protective devices for its own ends by placing the anemone on its shell. Why the sea anemone stays there and whether it gains anything from this relationship is not known.

Another crab, *Melia tessellata*, has taken this association one stage further, and carries a small anemone in each of its pincers. This must be one of the most ingenious cases of tool use by any animal. If approached by a predator, the crab confronts it with its outstretched claws bearing the stinging anemones, rather as primitive man might have held a wolf at bay by using a flaming torch.

Crabs normally catch food with their pincers. In the case of *Melia tessellata*, these pincers, which are specially adapted for holding the

*The crab,* Melia tessellata, *menacingly waves its two anemones to threaten its enemies.*

*Usually the hermit crab utilises a discarded whelk or winkle shell to protect it from predators, but when none is available a plastic jar will do.*

## HERMIT CRABS

The 800 species of hermit crab that occur in the world's oceans are distinguished from all other crabs by living inside the unoccupied shells of marine snails. These shells provide a home for the crabs and protect them from predators and the elements. Different sized crabs utilise different sized shells, and as a crab grows it must change shells regularly to accommodate its larger body. Hermit crabs must occupy shells of the right general size if they are to survive and breed. Several studies have shown that crabs in shells which were too small for them were much more likely to be eaten by predators.

In some areas the size of the shell is less important than the thickness of the shell. Other crabs and certain fish specialise in crushing mollusc shells, and where these predators exist, hermit crabs with the thickest shells are likely to be the safest.

Hermit crabs are able to make a rapid assessment of whether a particular shell is of an appropriate size or not, and how they do this has been the subject of much study. It appears that they use a variety of factors. The internal volume of the shell, shell weight, and aperture size are all important. Crabs that live in areas where empty shells are scarce select relatively large shells when given a free choice, compared with crabs living where there is an abundance of empty shells. Crabs that are short of shells presumably allow themselves room for growth. Suitably sized shells are often in short supply and in some areas this limits the hermit crab population. In one study in North America 12,000 empty shells were added to a small reef, resulting in a substantial increase in the crab population.

Where empty shells are difficult to find, hermit crabs often have to compete for them. Interestingly, this competition is highly ritualised. When two crabs meet in a rock pool, one of them grasps the other and 'raps' vigorously and noisily at the other's shell. Then, as though at some pre-arranged signal, both crabs wriggle free from their homes and exchange shells. Although aggression under these circumstances is ritualised and hence harmless, the situation is rather different if one of the crabs has no shell. When this happens the naked crab can sometimes evict the other by direct aggression.

An enforced exchange of shells was observed. In this case, there was an encounter between two individuals, one of which was sporting a rather battered shell. This particular crab grabbed the other individual and forced it out of its shell. The 'naked' crab moved about 60 cm (2 ft) away while the first crab nipped out of its damaged shell and took up residence in the new one. The other crab, rather than being left homeless and therefore extremely vulnerable to predators, instantly jumped into the broken shell. Clearly a damaged shell is better than no shell at all.

anemones without damaging them, are otherwise occupied. *Melia tessellata*, therefore, uses a pair of specially modified walking legs to pass food to its mouth, and is thus able to keep permanent hold of its two captives. Interestingly, even if the anemones are removed from the crab's claws, it still uses its walking legs for feeding. Fortunately, crabs have enough legs – ten in all – to be able to spare a couple.

Opposite *An archer
fish achieves a direct
hit.*

The diversity of form and behaviour among the several thousand species of
fish in the world is enormous, but there are probably just two, very
different, examples of tool use among them. The first concerns a small fish
*Butis* which occurs in the Cigenter river in Java and which was discovered
by the German wildlife photographer, Dieter Plage. He was sitting beside
the river watching dead leaves floating downstream when he noticed that
one leaf had suddenly started to behave very peculiarly – it was floating
upstream. Lunging forward with a net, he caught not just a leaf but a tiny
fish as well. When the two were placed in an aquarium the association
became clear. This small fish, new to science and a member of the family
called sleepers, swims upside down using some of its fins to hold onto the
leaf, and the other fins to propel it through the water. This behaviour is
unique among fish. Its function is not known as yet, but the leaf may serve
as mobile camouflage to protect the fish from the eight species of
kingfisher which feed along this river or it may act as a 'stalking horse' as
the fish pursues its prey.

There can be no doubt, however, about the function of the second case
of tool use among fish. The archer fish lives in mangrove swamps in
South-east Asia, where it frequently searches for food near the surface of
the water. When it spots an insect on overhanging vegetation, it lines itself
up and then shoots a jet of water from its mouth to knock the insect out of
the air. The archer fish is here using water as a tool, much as the ant-lion
uses sand, to help it capture its prey.

The archer fish possesses a number of structural and behavioural
adaptations associated with its ability to shoot down flies. The jet of water,
which can be projected over a metre, is produced by compression of the gill
covers. This forces water through a tube formed by the tongue pressed
against a groove in the roof of the mouth. When taking aim, the fish also
has to make allowance for refraction – the bending of light as it travels from
the air into the water. If it aimed at the point where the insect appeared to
be, it would miss. It therefore has to aim at a carefully judged angle to the
apparent direction if it is going to score a hit.

The fish possesses a further adaptation to enhance its accuracy:
binocular vision. The archer fish's large eyes can be aligned so that, like
our own, they produce a three-dimensional image, making it possible to
judge distances. Although it occurs among mammals and birds, few other
fish possess binocular vision.

The combined effect of these adaptations is to make the archer fish a
highly efficient predator of insects. When prey is about one metre away, the
fish's aim is almost perfect. At distances greater or less than this, accuracy
decreases. Immediately the archer fish has fired, it swims rapidly to the
point where its prey will fall into the water, ready to catch and eat it. Some
archer fish are able to line up their prey so precisely that it falls directly into
the water below, and within easy reach of the fish, instead of being knocked
inconveniently into the distance.

*The technical skill of the house martin enables it to construct gravity-defying mud nests beneath the eaves of houses.*

Some of the finest examples of technical skill are to be found among birds. The bowerbirds of Australia and South-east Asia, for example, construct fantastic bowers from sticks as a means of attracting the females. Most species decorate these avian palaces with flowers, mollusc shells and any stray human artefacts, such as clothes pegs, they can find. Three species, the satin, spotted and regent bowerbirds, also 'paint' their bowers with a mixture of charcoal, fruit and mud. The satin bowerbird uses a 'brush', usually a small flake of bark, to apply the paint to its bower; the other two species use their beaks.

Another ingenious construction technique is demonstrated by the house martin. In common with other members of the swallow family, house martins build elaborate mud nests precariously slung beneath the eaves of a house. By what magic do they get their nest to stay there? If you or I were to mix mud and water and place it in the same position it would fall off. However, if you watch house martins you will see that they do not simply 'plonk' their mud pellets onto the wall; instead, they place them in position while rapidly vibrating the head. This has an effect rather like that of a cement mixer. It creates a thixotrophic – or 'non-drip' – mixture which doesn't slither down the wall or fall off when put in position. This behaviour is almost too quick to see with the naked eye. It was described to us by an experienced wildlife film-maker, Alastair MacEwen, who was filming house martins with a high speed camera. When he played the film back at normal speed, which then makes the birds appear in slow motion, he could see what the birds were actually doing. As with other examples of apparent technical brilliance, it is natural selection that has produced this innate behaviour, rather than any genius on the part of the house martin. 'Cement mixing' has clear survival value – house martins that did not practise it and whose nests fell off the wall would have few surviving offspring. It has, therefore, through natural selection, become part of the behavioural repertoire of the house martins.

As well as their skills in construction, birds demonstrate considerable technical ingenuity in their efforts to get food. Several birds, including crows and gulls, carry shellfish into the air and drop them onto hard surfaces in order to break them open. Although this habit does not qualify as conventional tool use, the behaviour of the Egyptian vulture certainly does. For many years there were rumours of Egyptian vultures breaking open ostrich eggs by throwing stones at them at close range. The behaviour is now well documented. The vulture selects a rock, picks it up in its beak and, standing about half to one metre away, throws it at the egg with considerable force. The stones weigh 50 to 500 g (1.8 to 18 oz) and are propelled by a flick of the neck. About half the throws hit their target and it generally takes eight direct hits and about five minutes to crack the egg open. On some occasions, instead of throwing the stone, the vulture may use it to hammer the egg. This form of predation by Egyptian vultures is thought to be a significant source of egg loss for many ostriches.

The Egyptian vulture does not seem to use its throwing abilities against other animals. One was once seen to use a stone to kill a monitor lizard, but this was probably a mistake! The only bird that regularly directs missiles at other species is the Australian brush turkey, which kicks sand and stones at lizards when it is competing with them for food.

The most common tools used by birds in feeding, however, are not missiles but 'probes'. In Australia, the orange-winged sitella, a nuthatch-like bird, uses strips of wood or bark to winkle insect larvae out of crevices. Similar behaviour occurs in the American brown-headed nuthatch. But the bird that is best known for extracting insects from wood by means of a tool is the woodpecker finch of the Galapagos Islands. This is one of 14 species of bird referred to as Darwin's finches. During a visit to the Galapagos in 1835, Charles Darwin was fascinated by these birds, and they eventually

*A woodpecker finch teases out a grub with a twig.*

played a central role in developing his theory of evolution through natural selection. Surprisingly, Darwin made no mention of the woodpecker finch's remarkable tool-using behaviour, and it was not until 1919 that it was first described.

The woodpecker finch wields a cactus spine or specially prepared twig to insert into insect holes. Sometimes the prey is impaled, at other times the tool simply helps manouevre the larva into a position in which the bird can reach it with its beak. The woodpecker finch can even modify its tools to suit the situation; cactus spines may be cut down until they are the correct length, and twigs may have their side branches nipped off.

One other interesting example of feeding technology is the use of bait to catch prey. Bait fishing has been reported among pied kingfishers, sunbitterns (only in captivity) and black kites, but the best known instance occurs in the green heron. This is a geographically widespread species and bait fishing has been recorded among green herons in North America, Africa and Japan. What happens is this: the heron selects its bait, walks to the water's edge and waits until it spots a fish. It then throws the bait onto the water and as the fish moves towards it, the heron lunges forward and seizes its prey. The type of bait used varies: in North America the main items are bread, fish food pellets, feathers or mayflies. In Japan the herons also use berries, twigs, leaves and a range of insects including flies, cicadas and grasshoppers: in addition the Japanese green heron not only uses these naturally occurring baits, but also makes lures by breaking twigs into pieces of an appropriate size.

That this form of tool use is learned rather than inherited, and requires a reasonable amount of skill, is illustrated by the fact that juvenile herons are not good at it. This is because they select inappropriate bait, mainly twigs and leaves. Although adult herons sometimes catch fish using these particular baits, they are much more successful when using insects.

Birds, fish and invertebrates provide some fascinating instances of the use of tools, but it is only when we come to the mammals, and particularly the primates, that tool using begins to approach anything like its full potential. Apart from primates, some of the best known tool users among the mammals are the African and Asian elephants, whose prehensile trunk gives them a considerable advantage when it comes to holding and manipulating objects. Captive and wild elephants have been observed holding sticks in their trunks to scratch themselves and using leafy branches as gigantic fly swats. Their ability to throw things with the trunk is also well known; in zoos it is not unknown for elephants to hurl lumps of their own dung at keepers and visitors. Wild elephants, however, seem less likely to throw things. Their size alone is enough to intimidate any other animal that may threaten them.

Wild African and Asian elephants frequently throw dust, vegetation and water onto their bodies. The water helps to keep the elephant cool and clean; and the dust may encrust the animal's skin and provide an extra layer of protection from biting flies.

There are some instances of sophisticated tool use by elephants. One of these occurred when several animals encountered a wire fence which they

## ANTING

Most ants secrete noxious substances as a means of defence. Some squirt formic acid, while others exude foul-smelling substances from their anal glands. You only need to stir up the nest of a wood ant with a stick (not with your hand) to be able to smell them. Over 200 species of birds have been seen anointing themselves with these substances. The process involves using the ants as tools, and is achieved in one of two ways.

The bird picks up one or several ants in its bill and presses them against its feathers. While doing this the bird usually adopts a particular posture, with wing and tail spread, allowing access to the base of its feathers. The fluids from the ants are actually mixed with saliva and spread with a stroking movement onto the feathers. This is referred to as direct anting.

The other method, indirect anting, is less common. The bird simply sits on top of an ants' nest and allows the ants to climb all over it, squirting their defensive fluids as they do so. The bird adopts a characteristic squatting posture with its wings thrust forward to allow the ants access to the important feather tracts. Every so often the bird readjusts its position by shuffling its wings or tail, stirring up the ants as it does so.

*A starling anointing itself with the defensive fluids of an ant.*

The most likely explanation of anting is that it helps birds like the European jay to keep their feathers in good condition. The ants' fluids may act as insecticides and keep ectoparasites, such as feather lice, at bay. An alternative suggestion is that the substances produced by ants might supplement the birds' preen oil in some way.

were unable to push over. To surmount this problem, the elephants started tearing down trees, which they laid on the fence until it collapsed, allowing them to step over it. Wire fences are not something that wild elephants encounter every day, so this behaviour could hardly have been learned, still less inherited. It would seem that the elephants were able to understand the problem and work out a possible solution to it. And the solution worked.

The ground squirrel is another mammal that has adopted the practice of throwing things – primarily at its main predators, snakes. California ground squirrels will even start throwing sand at the mere sound of a rattlesnake.

Among carnivores, bears and the sea otter are the animals that most regularly use tools. Polar bears for example will occasionally throw lumps of ice or rocks at seals while hunting them, and in captivity they regularly throw objects around, apparently in play. Other bears will also use tools in captivity; spectacled bears use leafy branches, provided for them to eat, to knock fruit off overhanging branches or to retrieve floating bread from water.

The charismatic sea otter, which inhabits the coastal waters off the west coast of North America, is one of the best known tool users. These marine otters of California regularly open mollusc shells, whilst floating on their backs, by banging them against a rock balanced on their bellies. The usual sequence of events is for the otter to dive and then surface clutching both a

*Underwater, a Californian sea otter carries a rock between its paws, and a sea urchin on its belly. It will use the rock to break the urchin open.*

*Opposite The Indian elephant's ability to manipulate heavy objects has long been exploited by man.*

47

mollusc and a rock. Turning onto its back, the otter then places the stone on its chest or tummy. Holding the mussel with both forefeet, it begins pounding it on the rock until it is sufficiently broken to allow the otter to extract the flesh with its teeth.

As well as feeding on mussels, sea otters also take abalones. This has brought them into conflict with the fishing industry as the harvesting of abalone is commercially important. Since the abalone is rather like a large snail in having a protruding 'foot', it does not need to be opened in the same way as a mussel for the otter to eat it. However, the way in which otters detached abalones from the seabed was, for a long time, a topic of much speculation. Early observers who saw otters surfacing with abalones noticed that the abalone shells were often damaged, suggesting that they might have been removed by force. Eventually divers provided first hand evidence that sea otters use rocks as hammers under water to dislodge the abalones.

The sea otter's ability to carry, hold and manipulate stone tools is closely linked with two special adaptations. One is the ability to retract the claws on its front feet, making it much easier for the animal to manipulate objects in its 'hands'. The other is the possession of a flap of loose skin between each forelimb and the chest. These unique pouch-like structures undoubtedly help the sea otter to carry food and rocks from the seabed to the surface.

---

The use of tools among primates, the apes and monkeys, is widespread. This is probably due in part to the fact that so many people have gone looking for it. Primate tool use has been studied by anthropologists in the hope that it might tell us something about our own evolution; by psychologists trying to obtain a better understanding of problem solving and intelligence; and by animal behaviourists simply because the use of tools among wild primates is part of their natural repertoire of behaviour. In the academic literature about tool using, the primates definitely lead the field.

Apart from this, however, there is no doubt that primates are very well adapted for tool using because their hands are so well 'designed' for grasping objects. Like our own hands, those of most primates allow them to manipulate objects with a high degree of precision. They have the vital factor of the opposable thumb, without which it would be impossible to handle and investigate objects with any degree of manual dexterity. Because of this great advantage, the range of tool use among primates is, as we might expect, impressive.

Primates use tools in many of the same ways as other mammals – although they probably use them better. Many monkeys, for example, use branches, stones or even their own droppings as missiles directed at actual or potential predators. The apes – chimpanzees, gorillas and orang utans –

The long parent/ offspring association gives the young macaques plenty of time to learn from their elders. The macaques' delicate hands look remarkably human and are well adapted for grasping and holding objects.

Apes and monkeys, like humans, have an opposable thumb which enables them to handle and investigate objects. However, humans are distinguished from all other primates by their precision grip which enables them to manipulate objects with much greater control and accuracy.

49

*Baby chimpanzees soon learn to use their hands and feet to manipulate things to their advantage.*

are better adapted for throwing things than monkeys since the structure of their shoulders, combined with their upright posture, allows them to throw with considerable force. Monkeys also use sticks to reach food that may otherwise be out of reach, and they can use stones to help them open fruit and nuts. An amusing example of defensive tool use was provided by a Barbary macaque of North Africa and Gibraltar who, on being chased, climbed onto a roof and hurled tiles at his human pursuers.

What puts primate tool use into a class of its own, however, is that it does, in many cases, indicate intelligence. Perhaps one of the most remarkable uses of tools either of us has ever witnessed was the result of a remarkable piece of thinking – we use the word deliberately – by a Japanese macaque. An experiment was carried out with a mixed age group of macaques. A long plastic tube containing a tasty morsel of food in the middle was placed in their cage. The tube was more than twice the length of an adult macaque's arms. The monkeys were left to try to extract the food for themselves. After studying the situation for a few minutes the first macaque ran off and got a long stick. Within seconds it had managed to work some of the food to within an arm's length of the tube entrance. It then grabbed the food and quickly ate it before any of the others could steal it. But nearby there was a very large adult macaque who had worked out an even better solution. The adult seized a very small baby macaque, which could actually fit inside the tube. It then pushed the baby up the tube, keeping a firm grip of its leg. Once the youngster had grabbed what food remained, the adult pulled it out of the tube and took the food from it.

This same species of macaque has also been seen to wash rice and potatoes, provided by biologists, in the sea before eating them. Such behaviour not only cleaned any dirt off the food but also coated it in salt which may have added to its flavour.

The champion tool user in the animal kingdom, however, is the chimpanzee. Chimpanzees use tools more often and in a greater variety of ways than any other animal, except man. Much of what we now know about the tool-using behaviour of wild chimpanzees has come from the remarkable long-term studies carried out at the Gombe Stream Reserve in Tanzania by British primatologist Jane Goodall. Chimpanzees have been seen to use sticks to help them dig for roots, and as levers to break open ants' nests. They also use sticks or grass stems to probe for food, and in particular they use them to 'fish' for giant termites.

The soldiers in a termite colony respond aggressively to anything other than their own colony members entering the nest, so when a chimpanzee pushes a grass stem into the nest, they bite it. The chimpanzee extracts the stem with the termites conveniently attached by their jaws, and then pulls it either through its mouth or hand to remove them. Termite fishing is not as simple or straightforward as it sounds. It obviously requires a considerable degree of skill, not only in placing the tool into the nest and extracting the termites, but also in terms of selecting or constructing an appropriate tool. Stems for fishing are usually selected from nearby vegetation; they are generally about 1 cm (½ in) across and 65 cm (26 in) long. Chimpanzees younger than two years old do not attempt to fish although they do watch their mother doing it. By the time they are three years old, young chimps are attempting it themselves, but at this age they are still very inefficient, often using tools of the wrong size. However, a further year of practice and they are nearly as proficient as the adults. As well as catching termites in this way chimpanzees have also been seen 'fishing' in the nests of bees to extract honey.

Chimpanzees also use leaves as tools, either as a sponge to collect water or, as an aid to personal hygiene, to wipe sticky substances such as fruit juice from their bodies.

The orang utan, the red ape found in the rain forests of Borneo and Sumatra, shares some of the chimpanzee's tool-using abilities. It uses sticks to move objects, including food, and to probe for termites, and it uses leaves in the same way as the chimpanzee – it even uses a large leaf as an

*This orang utan demonstrates its ability to grasp efficiently with both hands and feet — a useful trick as it spends most of its life in the trees. This means it could hold on with a couple of limbs whilst feeding with the others.*

umbrella during rain. It does, however, spend more time than chimpanzees in trees, from which it occasionally drops branches onto predators.

Alone among the great apes, the gorilla seems to have a rather limited tool-using repertoire. In the wild, it confines itself mainly to unaimed throwing and the brandishing of sticks.

———————————— □ ————————————

Much of the fascination with tool use among animals stems from its apparent similarity to human use of tools. Because we think it so characteristic of human activity, we tend to assume that tool-using behaviour must demand special, exceptional intellectual skills. It was once thought that tool users might have a more complex and sophisticated nervous system than other animals, and that by studying them we might find out something of significance about the human race itself. Tool using was thought to be the 'key' to understanding ourselves.

It is now apparent that too much weight has been attached to tool using. To the animals that do it, there is nothing particularly special about using tools: it is a piece of behaviour much like any other that the animal performs. In some cases, such as the crab *Melia* and the sand wasps, tool use is innate, while in others, such as the green herons' bait fishing, it is largely learned. But whatever the case, tool use is only one of the many varied and complex behavioural adaptations that animals have acquired in their struggle to survive. Nevertheless, the topic remains a fascinating one: the magic of seeing animals like *Melia*, as it waves its two fearsome anemones at a predator, is not to be denied.

# BIGAMY BIRDS

A charming woodland scene: a male pied flycatcher, in his sparkling black and white summer plumage, singing his simple but pleasant song, has attracted a female to his nest site. She is building the nest while he jealously guards her against the attentions of other males. Soon, the female will begin to lay her eggs. The atmosphere is one of happy domesticity. This doesn't last long. As soon as the female begins to lay, the male leaves her; he returns to another part of the wood where another female – his first mate – is about to hatch her eggs, and takes up his parental duties. The abandoned female is often left to rear her family on her own.

The male pied flycatcher (as the chapter title suggests) is a bigamist. There is nothing so unusual about this. Although the great majority of birds are monogamous, there are plenty of species – particularly among the grouse, sandpipers, manakins, birds of paradise and cotingas – where it is common for males to have two or even more mates. What is different about the pied flycatcher, and what makes this species so interesting, is that the male achieves his bigamy by deceit.

The pied flycatcher is a summer visitor to Europe, having spent the winter months in Africa. The first birds, usually males, arrive in southern Britain in the second half of April, arriving a little later further north. Pied flycatchers breed primarily in broad-leafed woodland which provides the insect food, mainly caterpillars, they require to feed their young. In northern Europe pied flycatchers also breed in coniferous forests, but usually in fairly low numbers.

Holes in trees provide nest sites, but pied flycatchers will also use nest boxes. It is because they take to these artificial sites so readily that pied

flycatchers have been studied in such detail: breeding in a nest box allows nests to be checked and the adults to be caught and ringed much more easily than if they were in natural cavities. Pied flycatcher populations are often limited by the number of nesting sites available. Holes in trees are often scarce and the problem is made more acute by competition from other hole nesters like the great tit, redstart and nuthatch. The effect of simply providing nest boxes can be dramatic. In 1983, in one Welsh woodland where there were no nest boxes, three pairs of pied flycatchers nested. In the following year, 100 nest boxes were put up in the wood, with the result that, in 1985, no less than 60 pairs bred. The extra pairs had been there all the time; they had simply found nowhere to nest before. In some parts of Britain the breeding population of pied flycatchers can reach 120 pairs per sq km and in central Sweden as many as 200 pairs per sq km have been recorded.

The pied flycatcher's return to Europe is timed so that breeding coincides with the flush of spring insects, particularly caterpillars. It must time the laying of its eggs so that its chicks hatch when caterpillars are most abundant, and most palatable. Most pairs in Britain lay in the middle two weeks of May, with the eggs hatching 12 or 13 days later. The bulk of the chicks' diet consists of leaf-eating caterpillars from oak trees, particularly those of the green oak tortrix, and these are at their best in the few weeks after the leaf buds come out.

Defoliating caterpillars can cause severe damage to oak trees – indeed, trees sometimes have to put out a new set of leaves later in the summer. In an attempt to deter the caterpillars, the oaks produce distasteful chemicals called tannins in their leaves which make them less palatable. However, the tannins are least abundant in the younger leaves. Natural selection obviously favours the most successful caterpillars, and these are the ones that hatch out early enough to get their jaws into the young, tender, comparatively tannin-free leaves. One result of this is that the pied flycatcher must be quick off the mark early in the season in order to get the most nutritious caterpillars. Those caterpillars that feed on older leaves, with a high tannin content, are less nutritious for the chicks. The tannins affect the absorption by the birds' guts of certain essential nutrients causing low growth rates which are sometimes fatal. A key adaptation in pied flycatchers, therefore, is the timing of their breeding season. If they start too early there are not enough caterpillars to go round, but if they start too late the caterpillars can be toxic.

*A female pied flycatcher. This species is as much at home in an artificial nest box as its normal nest site – a hole in a tree.*

On their arrival at the breeding grounds, male pied flycatchers find a suitable nesting site. Unlike many other small birds, they do not defend a large territory, but simply defend an area around the nest hole. This behaviour, as we will see, plays an important role in the pied flycatcher's bigamous life. By means of his smart plumage and his simple song, the male tries to attract a female to his nest site. Once he has succeeded, the female enters the nest site and starts to build the nest. This, his first mate, is known as the primary female. The interval between initial pairing and the laying of the first egg is about nine days. During this time the nest is completed and the pair mate. Once egg laying has started, no further copulations take place. In the days preceding laying, when the female can be fertilised, the male remains close to his partner, guarding her from other males' attempts to mate surreptitiously.

The primary female lays a clutch of about seven eggs on average, laying one egg each day, generally in the morning. The number of eggs a female lays, however, largely depends upon when she starts. The earliest breeding

## SNEAKY MATINGS IN THE ZEBRA FINCH

The zebra finch is a small, colonial Australian finch, inhabiting all the drier parts of the country. It is a remarkable little bird in many respects, being superbly adapted to the unpredictability of a semi-desert environment. It can go without water for as long as 18 months, and is capable of breeding at just three months old if conditions are suitable.

Zebra finches are basically monogamous, with the male and female working together at all stages of the breeding cycle. However, males are sexual opportunists and if they come across another male's female they will try to mate with her. At the same time they try to make sure that no other males get the opportunity to mate with their own female. They do this by guarding the female, staying close to her during the ten days or so when she is fertile. This is not easy, however, because the male has to build the nest and guard his partner at the same time. The female sits inside the nest while the male goes off collecting nest material. He tries to confine his searching to the area around the nest, presumably so that he can keep an eye on his partner. If she leaves the nest to go and feed he always follows her, and his presence is usually sufficient to deter other males.

Despite his precautions, he sometimes loses track of her and she is left unattended. When this happens, the male flies around looking for her, calling frantically. Other males appear to be able to tell when females are near to laying, and try their best to seduce them using prolonged bouts of song and courtship display: sometimes they succeed and the female allows a strange male to mate with her.

Males have an even more direct way of achieving a 'sneaky' mating. When a pair of zebra finches are courting, they try to do so away from other colony members, and with good reason. Some males hang around the mating areas waiting for courting couples. Just as a male is about to mate with his partner, the bystanding male swoops in to take the other's place.

Detailed studies of zebra finches, both in the wild and in captivity, have shown that sneaky mating outside the pair bond can be successful in fertilising eggs. In general, though, it is the last male to mate that fathers most offspring. It is probably for this reason that male zebra finches copulate frequently with their partners, even though a single mating is quite sufficient to fertilise all the eggs.

*Zebra finches are always on the lookout for sex outside their 'marriage'.*

females lay the largest clutches and produce the most young, usually six or seven chicks. The last females to breed lay about six eggs but rear only four young. This seasonal decline in reproductive success is yet another factor in the pied flycatcher's bigamous mating system.

As soon as the female has laid her first egg, the male leaves her. He flies to some distant part of the forest – usually 150 m (nearly 200 yards), but sometimes as far as 3.5 km (over 2 miles), away. He finds another nest site and starts the whole procedure all over again. With luck, he succeeds in attracting and forming a bond with a second female. However, as soon as this female lays her eggs the male deserts her, leaving her to rear this brood by herself. He then returns to his first mate, the primary female, whose eggs are by now just hatching, and helps her raise this brood.

The mating behaviour of pied flycatchers is immensely complicated and scientists studying them interpret their behaviour slightly differently. This strange kind of polygyny has accounted for at least nine different hypotheses attempting to explain the birds' bizarre mating system.

Although the pied flycatcher's adultery has been known about since 1950, it is only in the last few years that the scientist Rauno Alatalo and his colleagues, studying pied flycatchers in Sweden, have begun to unravel the finer details. The observations reported below are based on the results of his studies.

The male deceives the second female by appearing to be unpaired. She pairs with him assuming that he will help her rear the young. The male pied flycatcher does nicely out of this arrangement: he fathers two broods, but normally helps to rear just one of them. On average, bigamous male pied flycatchers produce eight or nine offspring in a season, whereas monogamous ones produce only five or six. Females are less productive than this and secondary females do particularly badly out of the system. Primary females usually raise five or six young, while secondary females raise only three or four. Not only do secondary females produce smaller broods than primary females but, being alone, they also have to do twice as much work to rear their young, which saps their strength.

Males can afford to let their secondary female struggle on alone because a single bird can still rear some young. The pied flycatcher's nest site, being a hole nest, provides protection from both predators and the weather, enabling the female to leave the young for short periods while she forages for food.

For the male pied flycatcher, bigamy is obviously a successful strategy, but it also requires quite complex behavioural adaptations. One of the key things for the male is to time his departure from the first female correctly. If he leaves her as soon as she has accepted him, he increases his chances of getting a second female, but by leaving her unguarded he risks being cuckolded by another male. Males, therefore, generally wait until the female has started to lay her eggs before going off in search of a second mate, since females rarely mate again once laying has begun. This seems to be the optimal solution for the male. If he waited until his first female had completed laying, he would have missed opportunities for finding another female. By leaving when he does, he still gives himself the chance of a second female but he minimises the risk of being cuckolded. His strategy is the one most likely to guarantee the maximum number of his own offspring.

The question is: why do secondary females accept their status? The answer is that while they may not have the male's full attention they do often get a good territory with plenty of food. Since breeding success declines through the season, females simply cannot afford to waste time checking to see whether a particular male is paired or not. Every day of delay reduces their chances of successful breeding still further. Bigamous males simply exploit the fact that females are in a hurry. By moving away from his first mate, the male prevents the female from assessing his true status until it is too late for her to do anything about it.

Secondary females typically lay fewer eggs than either primary females or monogamously mated ones. This suggests that secondary females are somehow able to recognise the fact that they are not going to get any help from the male. Secondary females adjust their clutch size so that they will be better able to rear the brood. It will be in the female's best interests to lay just four eggs and rear all the young than to lay seven eggs and produce a brood of weaklings. This is better for the male, too, and he may actually inform the female in some way that she is a secondary female, but as yet we do not know how.

Bigamous males are usually larger than monogamous males. They may also be older and more experienced, and arrive on the breeding grounds earlier in the spring. Although all males try to get a second female, only about 15 per cent are successful. (This percentage can vary from year to year and fluctuates from 5–30 per cent.) Their success hinges on deceiving the female and, since this is achieved by having a separate territory for each female, it is easy to see how larger and more experienced birds might have the advantage.

A few other bird species also show this type of deceptive behaviour. These include the great reed warbler, the wheatear, the whitethroat and the redstart but, as yet, none of these is as well studied as the pied flycatcher.

———————————— □ ————————————

The male pied flycatcher is unusual in using deception to achieve his ends. He is far from alone in having more than one mate. Polygamy, the general term for this behaviour, is widespread in the bird world, and takes a bewildering variety of forms. It is quite common for male birds to mate with several females (technically known as polygyny, meaning many females). It is rather less common for females to mate with more than one

male (polyandry, meaning many males). And there are even species which do both (known as polygynandry).

The commonest form of polygyny occurs in species where the male obtains a territory and then attracts as many females to it as possible. The North American red-winged blackbird is a good example. The male of this starling-like bird is black with brilliant orange and red epaulettes. Redwings breed in marshy areas throughout the United States and much of Canada. Males arrive on the breeding grounds a day or two ahead of the females and set up territories. Once the females arrive, the male displays frantically in an attempt to persuade them that his own particular territory will be the best place to breed. The male's display involves a distinctive song and the flashing of his red wing patches. For some males this works well and they end up with as many as ten females on their territory. Others are less successful and may get only a single female or even remain unmated.

There are some puzzling aspects to this. Why should a female want to pair up with a male that already has several females when there are bachelors around? A female that pairs monogamously with a male will get some help from him with rearing the chicks. A female that moves in with an already paired male must share him with the other females. The females are obviously well aware of this, as they try hard to dissuade other females from joining their chosen male. Nonetheless, some males seem to be irresistible and end up as harem holders.

The reason for this appears to be that the territories occupied by redwinged blackbirds in these marshes vary enormously in their quality,

*The brilliant epaulettes of the male red-winged blackbird may be a help in attracting a large number of females to his territory.*

59

specifically in the amount of food they contain. Females first check out the males and their territories, and only pair with those in good territories. In some cases the territory is so good it doesn't matter that a female has to share it with other females. The male red-wings do quite well out of this arrangement since they father most of their female's offspring – most, not all, since the male cannot keep his eye on all his females simultaneously and some will happily mate with males in the territories next door.

The American lark bunting has a similar breeding system, but males mate with a maximum of two females. Lark buntings inhabit the prairies, breeding in alfalfa fields. Midday temperatures can be high and the major causes of chick mortality are sunstroke and overheating. Females prefer to mate with males whose territory contains a shady place to build the nest and rear the young. Males with little or no shade in their territory often remain unpaired. When biologists studying these birds added extra shade to some territories, the birds exhibited something that had not been seen in this species before; trigamy, males with three mates.

## DELAYED FERTILISATION

There is one simple, physiological reason why male birds try so hard to mate with other males' mates, and why some females let them. This is the fact that most, if not all, female birds store sperm after mating, before using it to fertilise their eggs.

Following mating, the sperm are stored in special tiny tubules in the female's reproductive tract. They may be retained there for days or weeks before swimming up to the top of the tract and fertilising an egg. The record, at present, lies with the turkey: some females have been reported laying fertile eggs ten weeks after their last mating. In other birds sperm can be stored for shorter periods: average values are about six days in quail and doves and about ten days in chickens, ducks and zebra finches. Compare this with most mammals, in which mating is timed to coincide much more precisely with the release of an ovum. Indeed, in some species such as rabbits, the very act of mating induces the female to ovulate. In most mammals sperm are viable in the female's reproductive tract for a matter of hours. In humans the record is about seven days, but this is exceptional.

The female's ability to store sperm has important consequences for birds' mating behaviour. It opens up the possibility for a male to come along and displace or remove the sperm

*The male roadrunner of North America offers his mate a dead mouse as a precursor to mating. Some species of birds mate frequently, others only once.*

stored from a previous mating. The female's sperm storage tubules are sausage-shaped and sperm lie at the bottom of the tube. Sperm from later matings overlie these, and are first out of the tubule when there is an egg to be fertilised: last in – first out. In other words the last male to mate usually has the best chance of fertilising the eggs. In birds, therefore, the fact that a female has already mated does not mean that the paternity of her offspring is assured. This is undoubtedly the reason why male birds often lead such frenetic sex lives. They are competing vigorously with each other to be last.

*The male red-winged blackbird attempts to attract as many females like this one, as possible to nest in his territory and, although the female would prefer his sole attention, she will share him with several other females if the territory is good enough.*

In some birds, females are attracted not by the quality of a male's territory, but by the male himself. The males of these species display in groups in order to attract females. The display ground is referred to as a lek, which, depending on the species, may be located in the centre of a large open meadow or deep in a tropical rain forest. The one thing all lekking species have in common is that, apart from mating with the female, the male bird plays no part whatsoever in raising the young.

Leks provide some of the world's most wonderful and memorable bird displays. Male blue-backed manakins of South America display in pairs, performing highly ritualised and marvellously synchronised displays. To attract a female, the two males sit side by side and call. Although both males call, they do so in such close unison that it sounds like a single call. Detailed analysis shows that the dominant member of the pair starts the calling, but the other male joins in just $1/20$ of a second later. When a female appears the two males start to perform the 'catherine wheel' display.

*The catherine wheel display of blue-backed manakins.*

The male nearest the female jumps into the air and, hovering, moves backwards. His place is then immediately taken by the other male. The birds go round and round, going faster and faster, and making nasal 'twanging' calls as they do so. Eventually the dominant male gives a sharp cry and the subordinate flies off. Left alone, the dominant male mates with the female who then goes off to lay two eggs and rear her young alone.

*Male black grouse or blackcocks congregate at traditional lekking sites where they compete for the attentions of the females. In Finland some of these sites are even found in the middle of frozen lakes.*

The cock-of-the-rock belongs to a group of birds known as the cotingas, closely related to the manakins. They too breed in the forests of Central and South America. The male cock-of-the-rock is one of the world's most colourful birds, with its brilliant orange plumage, black wings and a centurion-like crest. Males display in groups on the forest floor, clearing small areas – their 'courts' – of all leaves and twigs. The arrival of a female at the lek drives the males into a frenzy of display, with much showing off of

their remarkable crests. Females visit the lek only to mate, and they appear to be very particular about whom they choose. Some males are obviously more attractive to females than others and get the lion's share of the matings. The less attractive males, however, do not just sit on the sidelines. Often they will fly onto their rival's court just as he is mating, and disrupt the copulation. Observations of individually marked birds have shown that this disruptive behaviour is adaptive. Disruptive males cause the females to leave the lek temporarily; when they return, the likelihood that they will mate with such males is increased.

There are also lekking birds in the temperate regions of Europe and North America. They include the sage grouse of North America, and the black grouse, ruff and great snipe of northern Europe. In all these species the males compete for dominance on the lek, and the females mate mainly

*A sage grouse in full mating display will compete for the attentions of females on the lek.*

63

*A male long-tailed widow bird floats across his territory displaying his enormous tail. Experiments have shown that female long-tailed widow birds prefer males with the longest tails.*

*Opposite The resplendent tail of the male peacock has always intrigued evolutionary biologists and dazzled peahens. The enormous handicap such a tail gives a male may in itself be what attracts the female to the male. It may make the statement: 'I have survived despite having such a cumbersome tail so mate with me and your offspring will be as successful as me.'*

with the most dominant males. On a typical sage grouse lek, just 10 per cent of the males get over 75 per cent of all matings. In other words, some males are extraordinarily successful while others may never mate. This pattern is common to all lek species that have been studied. It results in intense competition between the males, and this may be one reason why the males of lekking birds are larger and much more brilliantly coloured than the females.

Quite why females should be so impressed by plumes, iridescent colours and whirling displays is an interesting question. It may be that these serve in some way as an indication of the quality of the male to the female. This idea has not been studied in detail in any lek species to date. However, some interesting experiments have been carried out with another polygamous bird, the long-tailed widow bird. This is a 7 cm (3 in) long member of the weaver bird family, the male of which bears a 50 cm (20 in) long tail. Widow birds live in Africa and, as in red-winged blackbirds, males try to attract as many females to their territory as possible. They do this by means of a beautiful display in which they appear to float across their territory showing off their enormous tails. A Swedish researcher, Malte Andersson, performed an ingenious experiment to see whether the male's tail had evolved because females showed a sexual preference for long-tailed males. He caught males and either shortened or lengthened their tails. The results showed that those males with experimentally elongated tails subsequently attracted more females than those with reduced tails. The females clearly preferred the males with the longest tails.

This study demonstrates that females do differentiate between males on the basis of their sexual ornaments. It is not entirely clear what females gain from this choice, but there are several ideas. One is that a long tail indicates an extremely fit male: to grow such a tail and carry it around will take extra energy. Therefore by choosing a long-tailed male the female will have her

eggs fathered by a strong male and such beneficial characteristics will be inherited by her sons. A second possibility is that the male's spectacular plumage indicates that he is free of parasites, otherwise he could not have grown such superb feathers. More research will have to be completed before we can determine precisely what excites female widow-birds.

In some polygynous species it is not the male's plumage that attracts the female, but specially constructed artefacts – bowers. The 18 species of bowerbird inhabit the damp forests of New Guinea and Australia. In one group of bowerbirds, the males are brightly coloured and build only modest bowers. In the other species, however, unlike their relatives, the birds of paradise, the sexes differ hardly at all and the males build some of the most extraordinary structures in the animal kingdom. The more complex the bower, it seems, the less elaborate the bird. The gardener bowerbird, for example, constructs the most remarkably complicated, hut-like bower, while the golden bowerbird builds a maypole bower up to 3 m (10 ft) high, gluing the sticks together with fungus and then decorating the whole structure with fruit, flowers and moss. Some species, like the satin bowerbird, construct avenue-shaped bowers, which are decorated with all sorts of items – snail shells, flowers, pebbles and an array of bits and pieces.

The bower is not a nest: its function is purely to attract the female. When a female approaches the bower, the male starts to display. Providing she is sufficiently impressed, she will mate with him inside the bower. The quality of the bower and its decoration play a key role in its effect on the female. As a result, males attempt to destroy each other's bowers, and steal each other's decorations. But, just as in lekking species, the males that put on the most impressive displays are the most successful in attracting females, while others with less appealing bowers may fail to mate at all.

*MacGregor's bowerbird builds a complex bower of the 'maypole' type. The bower is decorated with a variety of mosses, ferns and human artefacts. Then the male performs in his arena – mating with as many females as he can attract.*

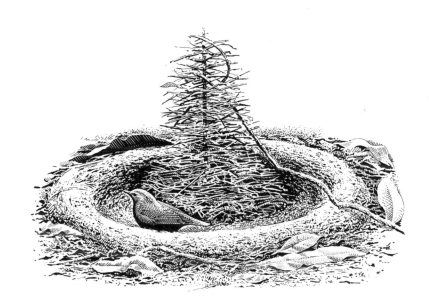

It used to be thought that a species was either monogamous or polygamous and that, in this respect at least, its behaviour was fixed. We now know that this is not true: in some species the mating system is quite flexible and changes according to ecological circumstances. The best known case occurs in one of Europe's most nondescript little birds, the dunnock. This sparrow-sized bird is dull brown and grey in colour, and since it spends much of its time scurrying around on the ground, it often looks more like a mouse than a bird. Despite being one of the commonest and most widespread birds in Britain, until recently it had not been studied in detail.

Our own interest in dunnocks was aroused several years ago by a chance comment, made over a cup of coffee. We were discussing dunnocks with a well known ornithologist when he told us that, as a schoolboy, he had colour-ringed the dunnocks in his garden and was surprised to find three different adults feeding a single clutch of young. This sounded interesting, so, working in a large garden in Scotland, we started to examine the mating system and social organisation of dunnocks for ourselves. We were indeed able to confirm that pairs of dunnocks were often joined by a third bird in

*The inconspicuous dunnock has the most variable mating system of any known bird.*

breeding, but this provided a mere foretaste of what was to come. It was not until Nick Davies, a researcher from Cambridge University, had studied dunnocks for several years that the awesome complexity of their love lives became apparent.

Dunnocks, it emerged, have the most variable mating system of any known bird. At one extreme some males simply remain unpaired because they cannot attract a female. Sometimes three males may share one female (polyandry) but two males sharing one female is the commonest situation. The next commonest is monogamy – one male paired with one female. But then things become rather chaotic. Up to three males may share two females, or two males may share three females. The commonest situation here is for two males to share two females. The final combination in this eccentric set-up is one male with two females (polygyny). What makes it all the more remarkable is that, unlike other group-living birds, the

## FEMALES WITH MANY MATES

Equality of the sexes has yet to arrive in the world of birds. The males of many species mate with several females, but females that mate with several males are rare. There are good reasons why this mating system, referred to as polyandry (literally, many males), is uncommon. The number of offspring a male can produce depends on his ability to produce sperm and inseminate females. Since sperm are manufactured by the million, the male's reproductive potential is really only limited by his access to females. The more females he can fertilise, the more offspring he will have. In contrast, a female's reproductive potential is limited by the number of eggs she can produce. Unlike sperm, eggs are produced very sparingly. So much energy is required for the formation of an egg that, in some species such as petrels, only a single egg can be laid in a season. Clearly, the number of males a female mates with has little or no effect on the numbers of her offspring.

There might, however, be other advantages to polyandry. If, by mating with more than one male, the female could persuade the males to help rear the young, then she would ensure a better chance of survival for her offspring than if she had only one helper or none at all. This does appear to be what happens.

Fewer than 50 bird species are polyandrous. These include jacanas, tinamous, phalaropes, some sandpipers and the dunnock. In some birds, like the American jacana, the female defends a large 'super-territory' within which each of her several males defends his territory. Curiously, the female will help each male defend his territory from a neighbour, even though a few minutes later she may well move into the neighbour's territory and switch sides. Female American jacanas may mate with up to four males at a time, sometimes within a few minutes, and then lay eggs in each male's nest. The result of this is that males have no idea as to whether or not they are the father of the offspring they raise. Nevertheless, there is still a sufficiently good chance that they are the father for them to undertake the major share of parental duties. One consequence of this role reversal is that in jacanas, as in many polyandrous species, the male is smaller and more cryptically coloured than the females. Since females compete for males, rather than vice versa, natural selection has favoured the females that are best at competing – usually the bigger ones.

The present-day breeding habits of some birds give us a reasonably good idea of how polyandry evolved. Female red-legged partridges and Temminck's stints produce two clutches in very rapid succession. The male looks after the first and the female the second. It is not too difficult to imagine how this could slip quietly into polyandry: females could develop the habit of mating with a second male and letting him care for her second clutch. In fact something exactly like this occurs in the American spotted sandpiper. The female mates with up to four males, laying a clutch of four eggs for each of individuals involved in these mating combinations are not close relatives.

The range of mating systems shown by the dunnock probably reflects the differences in the interests of males and females. It is in the interests of both sexes to maximise the amount of contact with the opposite sex. Females benefit from having several male partners because they all help to rear her young. A male benefits from having a number of females because he fathers all their chicks. Consequently, at one extreme, where one female has several males, she has 'won'. At the other extreme, where a single male mates with several females, he has 'won'. The combinations in between are those cases where neither sex has clearly won, but neither do that badly.

It is worth looking in more detail at one of the intermediate situations in this conflict between the sexes. The commonest of these is the one in which two males share one female. Each male 'wants' to be the sole father of the young, but the female 'wants' both of them to help rear her offspring.

them, but usually staying to help the last male.

Another form of polyandry is shown by the Galapagos hawk. Here again, a female might be paired to four males, but in this case there is just one nest and all the males help the female. However, the males compete vigorously for a share of paternity: if one male mates with the female the others all mate with her in rapid succession.

*Female liberation is the norm amongst jacanas like this pheasant-tailed jacana. One female mates with several males, leaving each to incubate a clutch of her eggs and raise the chicks.*

69

*The dunnock has a most unusual pre-copulatory display. The male places himself behind the female and pecks at her cloaca. Such behaviour may make the female deposit a drop of fluid which sometimes contains sperm from previous matings.*

As a result one male, the dominant of the two, does his very best to keep the subordinate from mating, while the female does her best to get both males to mate with her. The reason for this is simple. If both mate with her, they both help to rear the young, but if the subordinate does not mate, he doesn't help the female.

There is a further twist in the sexual politics of the dunnock. Before mating, the birds perform a display which is unique in the animal kingdom. The male pecks at the female's cloaca, or reproductive opening, and the female responds by ejecting a tiny droplet of fluid. On examination, it turns out that this fluid contains millions of sperm. By ejecting the sperm from previous matings, the female may convince the male that her offspring will also be his, and so enlist his aid in rearing them. This probably explains why male dunnocks copulate with such extraordinary frequency – about ten times a day. Effectively, they are competing to be last. The more often they mate, the more likely they are to be the one who fertilises the eggs, and the more certain they can be that the offspring they expend so much time and energy in rearing are theirs.

—————————————— □ ——————————————

Successful reproduction is obviously the prime objective of a bird's – indeed any animal's – life. But within this broad objective are a host of individual interests. Males compete with other males for the attentions of females; females then choose the most suitable males. Males guard their females against the attentions of other males; females attempt to prevent their males from straying in their affections. Within the bird world as a whole, these conflicting interests are resolved in a huge variety of ways, from the steadfast monogamy of the majority of birds to the bewildering polygamy of the dunnock, and from the resplendent tail feathers of the male widow-bird to the male pied flycatcher's deceit.

# THE SPADEFOOT TOAD

The Sonoran desert of Arizona is one of the least hospitable places on earth. The soil is baked dry by the fierce heat of the sun; there are no trees to provide shade. There may be little rain for ten months of the year – and in some years there may be no rain at all. It is hard to believe that anything can survive in such a hostile environment. Yet when the rains do finally arrive, the desert begins to stir. From burrows deep beneath the surface of the desert soil, small creatures start to appear. These are animals that would seem to have no chance of survival in the desert – they are toads.

Toads are amphibians and, like the other amphibians – frogs, newts and salamanders – they depend entirely on water for their survival. They cannot breed without water, and their thin skins must be kept moist or they will dehydrate and die. Anything less promising as a design for living in the desert is hard to imagine. Nevertheless, one group of toads, the spadefoot toads, have adapted to life in some of the hottest and driest places in the world.

There are several species of spadefoot toad; six can be found in North America, four in Europe, and several others in Asia. But the best known are the Arizona desert dwellers – the New Mexico spadefoot and Couch's spadefoot. All spadefoot toads have adapted to arid environments, but it is these two species that are most finely tuned to the harsh extremes of life in the Sonoran desert.

The desert environment is characterised by very low and unpredictable rainfall. When rain does occur, it is usually sudden, heavy and brief. Since rainfall controls plant growth, and plants are the base of the food pyramid, deserts typically have only a short period when food is abundant; it is during this time that the creatures of the desert must complete their breeding cycle. Most deserts are also extremely hot. In the Sonoran desert, average summer temperatures are around 40°C, and maximum temperatures may exceed 55°C. These are shade temperatures and do not reflect the real intensity of the heat that an animal living there might experience. Although air temperatures may be 40°C, the surface of the sand could be as hot as 70°C – hot enough to fry an egg.

*This spadefoot toad is at home in the dry desert lands of North America.*

Animals and plants living in the desert therefore face severe problems. They have had to adapt to high temperatures, shortage of water and a very brief time in which to reproduce. The fact that the spadefoot toad has achieved this is all the more remarkable because it is an amphibian, and its need for water is so much greater than that of other land-based animals. And the means by which it has done so provide another example of the fascination and complexity of adaptation in the natural world.

72

Amphibians are particularly vulnerable to hot and dry conditions because their skins are so delicate and permeable to water. Unprotected in the desert, they would lose so much water by evaporation that they would quickly die of desiccation. Most desert animals reduce water loss by finding shade in some form or other. In the case of the spadefoot, this means digging a burrow. They are well adapted for the task. All the spadefoots have a sharp-edged horny plate on the hind feet, the 'spade' from which they get their name. Using this 'spade', they simply dig down vertically until they are about 50 cm (20 in) deep. The burrow collapses around them so that they are effectively buried alive.

Within the burrow, the spadefoot achieves a high degree of protection from the relentless heat of the desert sand above. During June in the dry season the temperature on the desert surface can exceed 60°C, but 50 cm below the surface, where the spadefoot rests, the temperature is just 25°C, and remains almost constant day and night. Surprisingly, the burrow also provides the spadefoot with a source of water.

Deep underground, the spadefoot's thin skin comes into its own, enabling it to extract whatever moisture there is from the surrounding soil. It does this through a special adaptation that allows it to adjust the concentration of its blood. At high concentrations, the blood becomes a sort of chemical sponge, capable of drawing in whatever meagre supplies of water exist, even when the toad's body already contains much more water

*Sunrise in the Sonoran desert, Arizona.*

*Buried beneath the desert soil in its underground burrow, a spadefoot toad shelters from the intense heat, waiting for the all-too-infrequent rains to come.*

than the surrounding area of soil. The thin permeable skin allows moisture to flow into the animal rather than out of it. Even when the soil is virtually dry, spadefoot toads can remain underground for prolonged periods by maintaining their blood concentration at the correct level. Unsurprisingly the nature of the soil can affect the toad's ability to retain water. Clay soils, composed of very fine particles, have a much higher affinity for water than silty soils. As a result it is much more difficult for toads to extract water from clay.

Despite these adaptations, when the surrounding soil dries out during the long dry season, the toads may still lose considerable amounts of body water. However, the spadefoot's body tissues have the remarkable ability to withstand the loss of large amounts of water. Couch's spadefoot can lose up to 60 per cent of its body water without suffering any long term effects. When the time comes, it can recoup any water loss and rehydrate very rapidly, although not by drinking. Water is absorbed directly through the skin – no adult amphibians are known to drink. Rehydration can occur within four hours if the spadefoot is in water, but it can also take place, albeit more slowly, if it simply sits in or on damp soil. The skin of the toad's undersurface is specially adapted for this; it is thin, smooth and richly endowed with blood vessels, which transport the water to the rest of the body through the bloodstream. Several species of desert-living frogs and toads can withstand dehydration, but not to the same extent. Even those mammals best adapted to desert conditions, such as the camel, cannot begin to match this ability.

The spadefoot will spend months on end in its burrow, living off its fat supplies and conserving moisture. Almost all its activity is restricted to the two months of the year when it rains. The spadefoot has no fixed breeding season. It simply breeds when conditions are suitable, that is, immediately after heavy rain. If there is no rain there is no breeding. In southern California, rainfall is so infrequent that Couch's spadefoot may breed only once in two years. So until it rains, the spadefoot sits immobile in its burrow, an unsuspected living presence beneath the parched desert floor.

The sound of heavy rain drumming on the desert surface above is the signal for the spadefoot toads to begin to dig their way out of their burrows. Even now, however, they may have to wait. Despite the rain, daytime temperatures can still be lethal, so the spadefoot is strictly nocturnal, with relatively large eyes and cat-like vertical pupils. But as soon as it is dusk, the soil becomes alive with toads frantically digging their way to the surface. And within an hour or so there can be thousands of toads on the surface of the desert. The events that follow are spectacular: they have been graphically described by A. N. Bragg in his book *Gnomes of the Night: the Spadefoot Toads*, (Univ. of Pennsylvania Press, 1965):

> Two and one half inches of rain had fallen in about two hours, the sky clearing just before sundown. Air temperature was 15°C at 8.00 pm, just before complete darkness, when I heard the first voices in an extensively flooded field. At first I heard only a few calls; within ten minutes, at least twice as many animals were calling, and, by circling about the pool, I observed that a steady stream of spadefoots was coming from all directions to the breeding area. I caught and examined a good sample of these. All were males of a single species, the Plains Spadefoot (*Scaphiopus bombifrons*). These males entered the pool as they reached it and were soon calling as lustily as those first there. Within an hour hundreds of spadefoot males were bobbing like so many light balloons on the pool's surface, each every second or less giving voice to a single loud 'Wah!'.

The heavy desert rains soon produce shallow, temporary ponds. These rapidly become the scenes of intense activity, not just of the spadefoot toads, but also of the other animals that are tuned to a similar life cycle.

## DEFENCES AGAINST DEHYDRATION

Amphibians are particularly prone to dehydration because of their permeable skins. Those that live in dry climates have therefore found a variety of ways of avoiding water loss. Some, like the South American burrowing frog, encase themselves inside a cocoon within a burrow. The cocoon is formed by layers of parchment-like skin tissue, which completely envelops the frog, leaving only its nostrils exposed. All amphibians shed their skin from time to time, but because the burrowing frog remains immobile once it is inside its burrow and cocoon formation has begun, the successive layers of shed skin remain attached to each other. Laboratory studies of the South American burrowing frog showed that one layer of skin tissue was added daily for about 40 days to form the cocoon. Not surprisingly, the more layers of the cocoon there are, the greater the protection from dehydration.

Tree frogs of the genus *Phyllomedusa* display a novel adaptation for avoiding desiccation. These frogs live in the arid parts of South America and, as their name implies, they spend much of their lives in trees. They therefore risk dehydration during the day from the dry air. However, by covering themselves with a mixture of fats and waxes secreted by special glands in their skin, the frogs can avoid drying up. Using a series of elaborate grooming movements, they spread the mixture all over their bodies. When it dries, the frog has a covering which reduces evaporative water loss by 95 per cent.

Male spadefoots make their way to the ponds first and start to call. The chorus produced by hundreds of males can be heard over a mile away. This attracts females to the pond, but the calls also serve another purpose. Several different species of toad may be present in the pond at the same time. Each species has a distinctive call recognised only by others of the same species, which helps to prevent males mating with the wrong females. Couch's spadefoot sounds like an unhappy goat. The plains spadefoot's call resembles a duck's, while the New Mexico species makes a noise like a finger being rubbed over a balloon.

*A male Couch's spadefoot calls to attract a female to the pond. Its distinctive call is said to sound like an unhappy goat.*

Because the pools are limited in number, and because rain has stimulated all toads in the area to breed at the same time, the breeding ponds become extremely crowded. For the spadefoot, mating is an urgent business: they have not even stopped to feed, despite their months-long fast. There is literally no time to waste. The rains may disappear as suddenly as they arrived; the pond may dry out within a few days, and so the whole cycle of breeding activity must be completed in the shortest possible time. As a result, all spadefoot mating activity normally takes place on just one night. Some males return to the ponds on subsequent nights, but because there are few females around, very little happens.

Given the desperate urgency of the situation, and the fact that males usually outnumber females, it is not surprising that competition between males for females is intense. For the male spadefoot toad, unlike the males of many other species, it is not a question of how many females he can mate with, but whether he can mate with any at all. Just mating with a single female may be difficult enough and very few will have the opportunity to

Opposite *Even during the brief rains daytime temperatures can be lethal, so even now the spadefoot toad must wait for nightfall before emerging from its burrow.*

77

mate more than once a year. Males are seemingly indiscriminate and grasp anything roughly the right size, including other males.

Before they copulate, frogs and toads pair up in a 'pre-mating' embrace known as amplexus. The way in which the male holds the female in amplexus differs between species, but the spadefoot male holds his female around her waist. Pairs of toads in amplexus are easily caught and the sizes of male and female can be recorded and compared with sizes of unpaired individuals (most often just males) in the pond. Studies of amorous amphibians have revealed two different mating patterns. In the first, only the largest males obtain a female. In the second, male and female are matched for size, so large males form amplexus with large females, and small males with small females. The first pattern, demonstrated by tree frogs, for example, could arise because larger males displace smaller ones from females, or because females prefer larger males. In the second pattern, where pairs in amplexus are matched for size, the males may be better able to resist displacement by other males. Thus, as far as the spadefoot is concerned, it is quite possible that what appears at first sight to be a random collection of toads arbitrarily mating with anything and everything is actually an ordered process.

## LIFE IN DESERT POOLS

Spadefoot toads share their breeding pools with a number of other animals and plants. A single downpour can rapidly transform a lifeless, shallow depression in the desert into a pool throbbing with life. The first organisms, which appear within a few hours of the pool forming, are bacteria, as well as tiny single celled plants and animals called algae and protozoa respectively.

After just one or two days a variety of shrimp-like organisms appear. These include fairy shrimps, clam shrimps and tadpole shrimps. It may seem odd to find these aquatic animals appearing in the desert, but the eggs of these shrimps probably lie all over the desert, collecting in the depressions which ultimately become pools. These animals possess the remarkable ability to withstand intense heat and drought for prolonged periods of time. They do this mainly as eggs, and then hatch out and develop rapidly once it rains and a pool forms.

The eggs of most species are small, and although they do not contain a large supply of nutrients they remain viable for years at a time, mainly because the tiny embryo is in a resting state. In some species, the outer covering of the egg is waterproofed to prevent the embryo from drying out. The eggs of the brine shrimp, for example, have almost unbelievable waterproofing and heat resistance. In the laboratory their eggs can be heated to 98°C for 16 hours and still hatch out as normal. Indeed, in order to develop properly, brine shrimp eggs have to undergo a period of desiccation. The eggs of other shrimps, by contrast, are not waterproof, but possess thick, spongy coverings which protect them from bright sunlight and abrasion. These are important survival factors since the egg is the dispersal stage in the shrimp's life cycle. The eggs need to be well protected since they are carried on the wind, continually exposed to the sun's fierce rays, and blown across the sand.

When these desert shrimps reproduce, the eggs are protected and carried by the female, either in a special sac-like structure as in the brine shrimp, or in a special pouch as in the tadpole shrimp. In some species, only females appear to exist – males are either extremely rare or non-existent. Females can reproduce asexually, that is without fertilisation by a male. This allows for very rapid reproduction. The same pattern of asexual reproduction occurs in water fleas, aphids and even a few lizards.

Frogs and toads engage in amplexus for several reasons. Since fertilisation is external, with the male's sperm being shed over the eggs as they are laid, amplexus ensures that the male is in the right place at the right time to fertilise the eggs. In addition, amplexus allows a male to guard a female from other males until she is ready to lay her eggs. In the spadefoot, the intense competition among males for females means that some over-anxious males do not wait for the females to get into the pond but instead waylay them on the way to the breeding ground; the pair then arrive at the pond already in amplexus.

Within an hour or so of reaching the pond and becoming clasped in amplexus, females start to lay their eggs. As they do so the male arches his back and draws his sexual opening adjacent to that of the female. His sperm then covers the eggs as they are laid and fertilisation occurs within 10 to 15 minutes. The eggs are surrounded by a coating of jelly-like mucus which, when first laid, is very thin, allowing the male's sperm to penetrate. This mucus capsule swells rapidly on contact with water, protecting the egg from abrasion and fungal infection, while the outermost layer enables the eggs to be fastened on to a plant.

*A pair of* Atelopus *frogs in amplexus. Because males and females of this species encounter each other so infrequently, when they do meet they will remain stuck together for several months.*

79

Spadefoots are nocturnal, and their eggs must be laid and fertilised during the night when the pond is relatively cool. But by the middle of the following day, under the relentless desert sun, the temperature of the water can have reached 40°C. The spadefoot has a number of ways of helping the developing embryo to survive this particular ordeal. In the first few hours

## AMPHIBIAN BREEDING STRATEGIES

Frogs and toads can be divided into two breeding types, 'explosive breeders' and 'prolonged breeders'. In explosive breeding, all reproductive activity is packed into a few days, with the result that it is highly synchronised. It is typical of species that breed in short-lived pools, like the spadefoot toad. But explosive breeding also occurs in many North American, European and Asian species that breed in permanent pools. In these species mating occurs in early spring and rapid breeding may be an adaptation to avoid predation.

In prolonged breeding, on the other hand, mating takes place over much longer periods of time, or indeed throughout the year. It is characteristic of frogs and toads that live in more stable or predictable environments.

The breeding aggregations of explosive breeders can be extraordinarily dense. In most species, the males compete vigorously for potential mates, grasping virtually anything that comes in reach including inanimate objects and other males. If another male is grabbed he will give a release call. The over-zealous male then realises his mistake, lets him go and carries on searching for a female. Once a female is grabbed in amplexus the male is likely to have to defend her from other males, and fights lasting hours or days and involving several males have been witnessed in numerous species.

Prolonged breeders do not form such dense aggregations. Instead, males tend to be spaced out in territories, and they attract females to them by calling. Female frogs will readily approach loudspeakers broadcasting frog calls, demonstrating that it is the call alone that attracts them. Since females in these species become available to males over a period of weeks or months, the possibility exists for some males to mate with several females during the course of a breeding season. For example, in a study of the North American green frog, three out of 25

males mated between two and five times, and accounted for 47 per cent of all matings observed. In contrast, among the explosively breeding European common frog only one out of 33 males mated twice, and 16 did not mate at all.

The theory that amplexus is a form of 'mate guarding' by males, preventing others from mating with the same female, is confirmed by comparing the duration of amplexus in explosive breeding European common frog, only one out Among explosive breeders, competition between males for females is intense and amplexus may last for several days or even for the entire breeding period. In contrast, among prolonged breeders, amplexus lasts a few hours at most. Explosive breeders also show a further adaptation to male competition – the development of enlarged nuptial pads on their hands. These pads are not needed simply to hold on to the female, who remains passive. Instead, their function is to help the male prevent others from taking the female from him.

In the frog *Atelopus* the male and female remain in amplexus for several months. During this time the male is unable to feed properly and becomes severely emaciated. It is not clear why this species has such a long period of amplexus; it is not an explosive breeder. One suggestion is

*The nuptial pads on a male frog's feet enable him to grip on to the females tightly when mating. This helps him to secure the female from the advances of other males.*

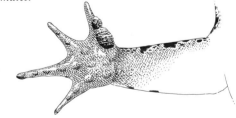

following fertilisation, the embryo is not particularly tolerant of high temperatures but, by the next day, its tolerance appears somehow to have increased. It is certainly helped by the fact that the spadefoot toad's eggs are black, covered by a pigment called melanin which helps protect them from the sun's rays (the same pigment is responsible for tanning in human

*Here the larger male bullfrog can take his pick from four females.*

that males and females rarely encounter each other. When they do, it is important that they literally stick together until they breed.

Within prolonged breeding species, individual frogs or toads have developed a variety of different reproductive strategies. The North American bullfrog demonstrates three of them. The first is the one adopted by the largest males: these defend territories in the pond and call to attract females who will lay their eggs there. These territories provide ideal conditions for the development of the eggs. The second is what biologists call the 'satellite strategy' and is the one followed by smaller males, who do not call but try to intercept females on their way to territory owners. The third strategy involves intermediate-sized males behaving opportunistically: they call from potential egg-laying sites but do not defend territories.

Alternative breeding strategies are also shown by many tree frogs in the genus *Hyla*, which typically call from positions that are not egg-laying sites. The two main strategies are 'caller' and 'satellite' and, just as in bullfrogs, the satellites are smaller, silent and try to waylay

females attracted to the callers. In some tree frog species certain individuals always call while others are always satellites. In other species individuals can switch strategies to become either caller or satellite depending on local conditions.

The European natterjack toad is a prolonged breeder. It is a small species with a distinctive yellow stripe down its back. Mating takes place at night in shallow pools in dunes and other sandy areas. Males sit in groups of three or four, the largest male calling from the centre of the group while the others act as silent satellites. In an experiment carried out by Anthony Arak, a zoologist from Cambridge University, the large calling male was removed from ten groups; in eight cases the other group members dispersed. In contrast, when a satellite was removed from nine other groups, in only one case did the others disperse. This strongly suggests that the peripheral males are exploiting the calling male's ability to attract females.

beings). In fact, most species of frogs and toads that lay their eggs in areas exposed to direct sunlight produce black eggs, whereas those species whose eggs develop in concealed places usually lay white eggs.

The way in which the female lays her eggs is also an important factor. Each female lays 2,000–3,000 eggs. These are laid in batches of several dozen and are then attached to submerged vegetation. Different species of frogs and toads deposit and distribute their eggs in different ways. Many frogs in the genus *Rana* produce large egg masses, whereas other species, like the spadefoot and many tree frogs, lay their eggs singly or in small groups, or in long strings, as in many other toads. Laying and distributing the eggs in small batches is a special adaptation to these hot desert conditions. Warm water holds less oxygen than cold water. By laying their eggs in small strips or bundles attached to vegetation, spadefoots ensure that each individual egg is, as far as possible, surrounded by water and thus receives the maximum supply of available oxygen.

The embryo has one further aid in its struggle for survival. The eggs develop extremely rapidly; within a remarkable 24 hours of being laid, the eggs will have hatched. The eggs of other frogs and toads not adapted to ephemeral pools may take as long as six weeks to hatch.

———————————————— □ ————————————————

The race against time continues. The spadefoot has one of the fastest rates of development of any frog or toad, but it still has to undergo the process known as metamorphosis – the egg does not hatch into an adult, but into a tadpole, which has to reach a certain critical size before it can finally metamorphose into a tiny toad. The pond in which the tadpole develops is only temporary, so it must complete its metamorphosis into a toad as quickly as possible. This crucial transition can take tadpoles of other species – those that develop in cold mountain streams, for example – two or even three years to complete. The spadefoot toad can manage the whole process, from egg to tadpole to toad, in as little as ten days. The rapid development of the spadefoot tadpole is a feature shared by other frogs and toads that occupy temporary ponds, such as the European natterjack toad.

The rate of development of spadefoot tadpoles depends upon several factors, including water temperature and the number of other tadpoles around. The number of tadpoles in a particular pond can have a dramatic effect on the food supply. If the density of tadpoles is high, then food will be scarce. As a result, development takes longer, a greater proportion of individuals die, and the survivors may actually be forced to change into toads at a smaller size.

The tadpoles of some spadefoot toads are gregarious and form large moving shoals. Among the tadpoles of the eastern spadefoot, shoaling occurs only when the water level in the pond is decreasing. The appearance of these shoals is striking for they consist of tadpoles of identical size. Such social aggregations may be either a feeding adaptation

or a means of avoiding predation. Tadpoles normally feed on plankton and decaying plant material suspended in the water and shoals may effectively stir up the debris at the bottom of the pond, thus adding to the food supply. In the plains spadefoot, however, the primary reason for shoaling is to avoid being eaten, since shoaling occurs only in the presence of such predators as snakes and the larvae of carnivorous water beetles.

As the waters of the pond slowly evaporate, crowding the tadpoles into an ever smaller volume and reducing still further the available food supply, an extraordinary transformation may take place. One of the spadefoot's most remarkable adaptations is the tadpole's ability to change both its appearance and behaviour to become a cannibal. In the western spadefoot of Europe, for example, there are two distinct forms of tadpole: a 'standard' type which feeds on rotting vegetation, and a cannibalistic type which kills and eats other tadpoles. Those tadpoles that turn into cannibals become larger, and develop a grotesquely shaped head with an enormous predatory mouth – a change in appearance more appropriate to a horror film than a television natural history programme. The extent to which spadefoot tadpoles become cannibalistic varies between the different species and according to ecological circumstance. The tadpoles of Couch's spadefoot will attack weakened and partially metamorphosed individuals when the pond begins to dry out, but will not attack healthy tadpoles. In other species, a proportion of cannibals is present whatever the circumstances, and will prey on minute shrimps and other freshwater animals as well as on their fellow tadpoles.

The gruesome ability of some individuals to take up a cannibalistic diet is, in evolutionary terms, a sensible solution to the problem of food shortage in a rapidly disappearing pond. The spadefoot has, in fact,

*These cannibalistic tadpoles make short work of one of their siblings.*

83

adopted a flexible strategy that allows it to cope with all eventualities. It is known that both cannibalistic and non-cannibalistic individuals can and do develop from the same batch of eggs. If the tadpole's normal food, decaying plant material, is abundant, then both types of tadpole develop and both types will survive to become adult. But if this food is in short supply and the whole population is in danger of dying out, then cannibalism makes sense: it maximises the food supply and vastly increases the chance that some at least of the tadpoles will survive.

Like all adaptations, cannibalism must have first arisen by chance but, in the harsh environment of the desert, it is easy to see how this could have happened. If a mutant batch of tadpoles with cannibalistic tendencies suddenly arose in a spadefoot population, they would tend to survive much better on those occasions when the pond was quickly drying out. As a result, the genes for cannibalism would remain in the population. Of course, it would be self-defeating if all tadpoles were to become cannibals, since none of them would then eat the plant material on which the chain depends.

While the idea of cannibalism may seem repugnant to us, it clearly has its advantages, and it is certainly not unique to spadefoot toads. It also occurs in other amphibians that develop in crowded temporary ponds, for example the South American bullfrog. There are also several poison arrow frogs of South America that show a rather different form of cannibalism. These small frogs inhabit tropical rain forest and lay their eggs in the tiny 'ponds' in the centre of bromeliad plants. A single fertile egg is laid in each plant and each tadpole develops in the privacy of its own pool. Not surprisingly, the food supply in such tiny pools is extremely limited. To get round this problem, the tadpole's mother provides it with an unusual food parcel – a supply of unfertilised eggs. Obviously this is not true cannibalism as in the spadefoots, but it provides an example of how other frogs have found a parallel solution to the problem of rearing young when food is scarce.

———————————— □ ————————————

Once the spadefoot's metamorphosis is complete the surviving toads leave the water. Sometimes, as in the eastern spadefoot of Europe and south-west Asia, tadpoles form large groups during the final phase of their development. As a result, vast numbers of tiny toads complete their metamorphosis and leave the pond within the same ten minute period at night – a scene every bit as dramatic as the mass migrations of wildebeest in the Serengeti, but in miniature. Obviously, since mating itself takes place mainly on the one night, and the development period is so compressed, one would expect the toadlets to emerge more or less together. But such a high degree of synchrony is extraordinary and has not yet been explained. It is possible that it has to do with cannibalism. Since cannibalistic tadpoles feed preferentially on partially metamorphosed or weaker individuals, there would be a tendency for those likely to hatch earlier or later to be weeded out. Even this does not seem a sufficient explanation. Perhaps this synchronous transformation is the means by which the toads limit predation by their own kind. Whatever the case, the mass emergence of spadefoot toads into the night is an unforgettable experience.

The young toads must continue to feed and grow before conditions become too dry again, but their battle for survival is not yet over. At this stage immediately after metamorphosis, they are vulnerable to another unexpected source of predation. In the mud around the edge of the pools, the larvae of a horsefly are developing. These too need a rich food supply to complete their development, and the newly metamorphosed toads, just a few millimetres long, are exactly the right size. The horsefly larvae use their two large jaws to capture the tiny toad and pull it down into the mud where they feed on its body fluids.

Even if they escape the attentions of the horsefly larvae, the young toads are still not safe. Both young and fully grown adults are a favourite food for predatory desert animals, mainly snakes and the hognose snake in particular. Spadefoot toads have a number of anti-predator tactics, one of which is to inflate the lungs and hence the body, which may fool the snake into thinking the toad is larger than it really is and act as a deterrent.

If this doesn't work, the spadefoot has a second line of defence; it has a disagreeable odour and taste – the New Mexico spadefoot has the curious distinction of smelling like unroasted peanuts. Some species of frogs and toads even produce intensely toxic substances in their skin. The deadly secretions in the skin of the poison arrow frogs of South and Central America are described in Venomous Animals (see page 99), but several

*A spadefoot toadlet leaving the desert pond in search of food. Its tail will shrink within a few hours of departure.*

# SURVIVING WITHOUT WATER

Most desert-living animals can withstand some water loss, but none can match the spadefoot and its desert-living cousins. Small desert rodents and the Bedouin goat can survive short periods of dehydration which result in a 20-30 per cent weight loss. Among mammals, however, the desert camel shows the greatest tolerance to dehydration. Whereas most large mammals can survive no more than 14 per cent water loss, the camel can survive an amazing 30 per cent loss for over a week quite easily. It does this by losing water from the reserves in its stomach and from the body tissues. Some vertebrates, including ourselves, lose water from the blood. This impairs the circulation which then means that the body cannot lose excess heat. The result is a dramatic increase in body temperature, until at 43°C death occurs. Camels can recover rapidly from dehydration by drinking sometimes vast quantities of water. Camels that have lost 20 per cent of their weight through dehydration can regain this in just ten minutes of drinking. One dehydrated camel drank 186 litres of water in two bouts of 94 and 92 litres.

The lethal limit of dehydration in humans is equivalent to about 18–20 per cent of body weight. One man was lost in the Mexican desert for eight days, having had only enough water with him for one day. The man survived, and this appears to be the record for a human being.

*Most of the legendary tales about the camel's ability to survive for long periods without water are true. This nomad is leading his camels to a water hole after five dry days in the desert.*

species of toads produce cardiotoxic steroids in the skin and, in one species, the secretions are hallucinogenic.

Those young adults that survive all these various hazards join their parents in the important business of eating. Feeding takes place at night and toads continue to forage for as long as conditions are reasonably humid, sometimes for as long as two months. During the day they retreat into shallow burrows a few centimetres below the ground. Their food consists of virtually any invertebrate small enough to swallow, including grasshoppers, spiders and caterpillars.

Left *A spadefoot toad falls foul of one of its predators – the hognose snake.*

As with every other aspect of the desert life cycle, time is of the essence. Soon the dry season will return and food supplies will disappear. Before then, the toads have to lay in sufficient stores of energy to sustain them during their patient vigil underground. Once the rainy season has ended, the surviving toads burrow once more deep into the desert soil. They will remain there in a state of torpor, patiently awaiting the return of the rains. But they may have to wait for a year, and maybe even two, before the brief, dramatic cycle can begin all over again.

Below *A bullfrog from Kenya, East Africa, consuming a mouse. The spadefoot toad feeds on smaller prey such as grasshoppers, spiders and caterpillars.*

# Venomous Animals

In the heart of the African rain forest a chameleon crouches on a branch, stalking an insect. This arboreal lizard is a formidable predator. It has binocular vision for locating its prey and now rests motionless, waiting for the potential victim to come closer. The chameleon will then dart out its long, sticky-tipped tongue, impale the insect and swallow it. On this occasion, however, the hunter becomes the hunted. So intense is the chameleon's concentration that it is quite unaware of imminent danger. Slithering unseen and unheard through the branches, a green mamba is itself planning to make a kill. The moment it is within range it strikes with sudden speed, burying its fangs in the soft neck tissues of its prey. The snake's venom courses through the bloodstream of the chameleon, causing instant paralysis. In a matter of seconds it is dead.

Green and black mambas are among the most venomous snakes in the world; and for most people, snakes, of all animals, have the most sinister reputation. Snake bites cause the deaths of over 100,000 people every year. What is not so generally realised is that many deaths and serious illnesses also result from the poisonous substances produced by a range of other animals.

Although 'venom' and 'poison' are often used indiscriminately to describe toxic substances, we could make the distinction here between them. Animals that are venomous secrete toxic matter and inject it by biting or stinging or by some other means. Poisonous animals secrete lethal toxins but do not actively attack other creatures with them. However, they can equally make people or other animals extremely ill, perhaps fatally so, if they are eaten. Despite the distinction, the results can be similar.

Opposite *A prairie rattler. Snake bites cause the death of over 100,000 people every year.*

*A Portuguese man-o'-war blown ashore on the Florida coast. The dangerous stinging tentacles hang down below the float.*

Poisons and venoms are important survival factors for many creatures. Some animals, such as venomous snakes and spiders, inject venoms in order to immobilise and kill their prey. Others employ them essentially in self-defence; the honey bee, for example, uses its sting to protect its colony. Sometimes, however, the distinction is less clear cut: thus the spitting cobra's venom serves both to subdue prey and as a means of self-protection. And among those animals which secrete poisons but have no means of injecting them, their function is usually defensive: in the case of certain tropical sea-urchins, it makes the eggs distasteful to predators.

———————— □ ————————

Many venomous creatures, in appearance innocent and harmless, inhabit the oceans. The Portuguese man-o'-war looks like a single organism but is in fact composed of numerous individual polyps, which resemble sea-anemones. Some of these are responsible for capturing and digesting food, while others form the gas-filled blue balloon which allows the animal to float on the sea's surface. Each of the many feeding individuals in the colony possesses an extremely long tentacle which bears large numbers of stinging cells known as nematocysts. Each stinging cell consists of a venom capsule and a reversible, barbed sting, evolved in order to capture prey such as surface-feeding fish. When a mackerel or a flying fish, for example, inadvertently grazes a tentacle, hundreds of nematocysts fire and inject their venom into the victim, which is quickly paralysed and killed. The fish is then passed up into the mouths of the feeding individuals, and the ingested food is shared among all the other (non-feeding) members of the colony through a series of cavities.

If a human swimmer collides with a Portuguese man-o'-war, the consequences can be very dangerous. Severe pain, muscle seizure and heart failure may end in rapid death; or the pain may be so intense that the victim is unable to swim and drowns. However, if he or she survives the first half hour, there is a good chance of complete recovery.

*An underwater view of a jellyfish – this is the aptly named sea nettle. Though not as deadly as the sea wasp, this jellyfish can cause severe irritation if it stings a human.*

The little-known sea wasp, in spite of its name, is actually a jellyfish. Like the Portuguese man-o'-war, it is equipped with nematocysts, used for catching prey. Its feeding technique is also similar: drifting downward through the water, it captures small fish or plankton with its tentacles. A tropical species, the sea wasp occurs around the coast of Australia. Measuring about 20 cm (8 in) across, but with tentacles over 1 m (3¼ ft) long, it is translucent light blue and difficult to see. It is much more dangerous to humans than the man-o'-war, and several hundred fatalities have been recorded; death is so sudden that no treatment is possible.

Relatively few molluscs – snails and their relatives – are venomous, although some are poisonous when eaten because they accumulate toxic substances through feeding. Mussels use their gills to filter tiny prey out of the water. During the summer these may include small animals known as dinoflagellates, which produce toxic waste products. If the mussel is eaten by an animal or by a person, the poisons that have built up inside the mollusc may prove fatal. About 8 per cent of all known cases involving human beings result in death.

There are two types of mollusc that are actively venomous: the cone shells and the blue-ringed octopus. Cone shells are marine snails of tropical waters with specialised predatory habits. Unlike most snails, which are harmless and graze on plant material, the cones possess a long feeding 'nose', the proboscis, which bears barbed teeth and venom. The proboscis can be shot out forcibly, stabbing its prey. Different species of cones specialise in different types of prey, mainly comprising worms, other molluscs and fish, and have specific venoms to deal with each of their prey. The victim is usually speared with the barbed teeth in the proboscis and quickly killed by the venom which is injected at the same time. The prey is then drawn up to the mouth and swallowed whole.

The venom from worm-killing cones is harmless to humans, and that of the mollusc-feeding cones seldom has serious consequences. But a bite from a fish-feeding cone can be both painful and dangerous. Some of these cones are quite large, with teeth 2 cm (almost 1 in) long; the fast-acting venom causes paralysis, breathing difficulties and sometimes death.

Many sea-urchins of temperate waters have tiny pronged pincers, often provided with poison glands. The poison is seldom dangerous but anyone who has accidentally trodden on a sea-urchin while bathing knows how painful the swelling can be. In tropical oceans there are some species whose reproductive organs are toxic, making its eggs distasteful to predators, it is thought. Unfortunately, the insides of these sea-urchins are considered by some peoples to be a delicacy eaten raw and so are potentially dangerous. Although the symptoms are extremely unpleasant (nausea, diarrhoea, vomiting and headache), few people have died.

Australia is host to some of the world's most dangerous animals. One of these is the infamous blue-ringed octopus. This, too, is a mollusc, closely related to squids and cuttlefish, and together they comprise a group of animals known as cephalopods. This literally means 'head-foot', for the head is situated next to the numerous legs, or sucker-bearing tentacles. All cephalopods are carnivores, feeding on fish, shrimps and crabs, which they catch with the tentacles. Prey is immediately carried to the mouth and killed with a bite from the bird-like beak, tucked away among the tentacles.

The blue-ringed octopus, which lives in the subtropical and temperate shallow waters off eastern Australia, is comparatively small, with a diameter of only 20 cm (8 in). It is dark brown, but attractively decorated with circles of electric blue. Like many cephalopods, it has a curious life cycle:

after growing very rapidly, reaching sexual maturity in just four months, it dies after a single breeding session.

The octopus kills its prey, mainly small crabs, by biting them and injecting highly toxic saliva into the wound. If handled or disturbed, the blue-ringed octopus may also attack a human, and although the bite is painless the symptoms are dramatic: vomiting, temporary blindness and lack of muscle coordination. Death usually results within two to three hours of the bite. All fatalities associated with this species have occurred when swimmers or divers have casually picked up and handled the octopus.

There are about 30,000 species of fish in the world and some 250 of these are actively venomous, possessing spines and venom glands. In almost all cases this venom is used as a means of defence, and most people are stung after handling or stepping on a fish. The principal venomous species include the scorpion fish, stingrays, weaverfish, rockfish and waspfish. However some fish, such as the puffer fish, produce poisons as a by-product of their natural metabolisms.

The puffer fish is found in warmer parts of the world both in sea water and fresh water. It has the strange ability to blow itself up with either air or water as a means of self-defence. The poison, which is a naturally occurring by-product, is found mainly in the liver, intestine and reproductive system and finds its way into the eggs. The fish is dangerous if eaten from May to July just before spawning. About 60 per cent of people who are poisoned by puffer fish die – usually within six to twenty-four hours – from respiratory paralysis. There is no known antidote to this poison. Astonishingly, the fish is considered a delicacy in Japan where it is known as 'fugu'; but because of its toxicity it can only be prepared by licensed chefs.

*Puffer fish, such as this spotted species from the Cigenter river in Java, Indonesia, could poison you if you ate one during its spawning season.*

93

*The splendid-looking scorpion fish can be found around coral reefs, but you are well advised to give them a wide berth as their venom can be fatal.*

The scorpion fish, so named because the effect of its sting is much like that of a scorpion, is a dramatically beautiful reef fish, boldly striped in red, pink and white, measuring 10–15 cm (4–6 in) in length, with greatly elongated fins. Almost all the major fins possess spines and venom glands. For anyone rash enough to seize it, the effect is instantaneous – searing pain, swelling, convulsions, unconsciousness and sometimes death.

In contrast to the scorpion fish, which flaunts itself, the stonefish and sculpins, which occur throughout the world, are superbly camouflaged, resting on the sea bed waiting for unsuspecting prey to swim past. The spines are situated on the dorsal fin and around the head and it is all too easy to step on them by mistake. The symptoms of poisoning are similar to those inflicted by the sting of a scorpion fish. Accidents involving bathers are especially common in the waters off north-eastern Australia, but happily an effective anti-venin is now available.

———————————— □ ————————————

The group of jointed animals known as arthropods includes scorpions and spiders, and almost all of these are venomous. Scorpions occur in the warmer parts of the world and are nocturnal by habit, hiding by day in holes or under stones. Most species measure 2–3 cm (about 1 in); the largest ever recorded, at over 30 cm (12 in), is now mercifully extinct.

The elongated body of the scorpion is armed at one end with two

powerful claws and at the other with a long tail terminating in a sting. Scorpions are carnivorous and feed mainly on insects which they catch with their claws and then kill by stinging. The stinging apparatus consists of a sharp barb and a bulbous base which contains the venom-producing glands. These glands are surrounded by sheets of muscle which, when contracted, squeeze the venom out into the barb. Males and females are similar in appearance, although the males have slightly longer claws.

Scorpions are much feared, and with justification. Today an estimated 150,000 people are stung each year but less than 1 per cent of these attacks prove fatal. Almost half of the incidents occur in Mexico, where a particularly dangerous scorpion, *Centruroides*, is common. Scorpions seldom go out of their way to attack people and accidents usually happen when scorpions are trodden on. However, not all scorpions are venomous and the same species in different parts of the world differ in their toxicity. For example, a sting from *Buthus occitanus* is likely to be fatal in North Africa, but in France the species, for reasons unexplained, is completely harmless.

Spiders are close relatives of scorpions and many of them can deliver a painful bite. The spider's mouthparts comprise two jaw-like chelicerae, each tipped with a sharp fang. The fangs are hollow and are linked to a venom gland. When a spider bites its prey, the fangs are driven into the victim and at the same time the venom is discharged from the gland, down

*Scorpions will only sting a person if threatened or if accidentally trodden on.*

95

the fangs and into the wound. The venom of most spiders, though effective against their insect prey, is harmless to man. Only a handful of species are dangerous. These include the black widow spiders, found in North and South America, Africa and many parts of Australia, and, to a lesser extent, tarantulas.

Spider venoms fall into two main categories: those which affect the nervous system (neurotoxins) and those that produce a necrosis, or death of the body tissues. The black widow spider and its relatives produce neurotoxins. Although these spiders are rarely aggressive and generally bite humans only in self-defence, farm workers are particularly at risk; and in some areas with outside lavatories, such spiders are known to lurk under the seats. The bite itself may be painless and even go unnoticed, but within ten minutes to one hour symptoms start to appear; a dull pain spreads from the region of the bite and gradually becomes more severe, causing cramps, vomiting, perspiration and copious salivation. In most cases, these symptoms persist for a couple of days. An anti-venin is available, however, and the majority of victims survive; less than 5 per cent of people bitten by black widow spiders die and in the United States only one or two deaths occur each year.

Rudyard Kipling's observation that 'the female of the species is more deadly than the male' is true in the case of the black widow spider. The males are small and harmless to humans, but the female is larger with a black abdomen bearing a bright red mark – a useful warning sign.

*The 13 cm (5 in) long Costa Rican wolf spider feeds on frogs. It kills its victims with venomous fangs. If you touch a wolf spider with your finger the sharp spines on top of the body and the legs will cause you searing pain.*

Centipedes, like spiders, kill their prey by injecting venom into them with a pair of large fangs, each of which is equipped with a venom gland. As a rule the victim is an insect, but certain large centipedes – and they may measure up to 30 cm (12 in) long – have been seen to capture toads and mice. Some of the large species, if picked up or trodden upon, can give a painful bite; but none are really dangerous and few human fatalities have resulted.

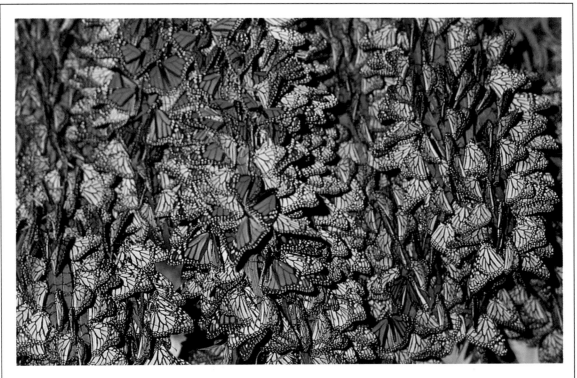

## THE MONARCH BUTTERFLY

*Both the caterpillar and adult of the monarch butterfly are distasteful to their enemies. They obtain their unpleasant flavour from the milkweed plant.*

The monarch butterfly, found over much of North America, has been much studied because of its noxious qualities. The females lay their eggs on milkweed and the caterpillars feed on these plants until they pupate, prior to emerging as butterflies. As protection against herbivores such as caterpillars, the milkweed produces chemical deterrents on its leaves. These substances, referred to as cardiac glycosides, have no adverse effect on the monarch caterpillar and it can accumulate and store these chemicals for its own use. The glycosides are even retained as the caterpillar changes into a butterfly and help to protect it from predatory birds. It is especially interesting that the highest concentrations of the toxins occur in the butterfly's wings rather than in its body. The advantage for the butterfly is that if a bird seizes it by the wings, and is reminded of some earlier unpleasant experience with the species and its toxin, it will release the insect unharmed: a bluejay that has never seen a monarch butterfly will readily catch and eat it; but within 15–30 minutes the toxins in the butterfly's body cause the bird to vomit, and one such encounter is sufficient to make a bluejay avoid monarchs thereafter. In addition to being distasteful, the monarch is warningly coloured with black and orange stripes, should the bluejay need another reminder.

Butterflies and moths are among the most beautiful insects and one does not expect them to be venomous or poisonous. Yet the caterpillars of a few species are well protected from predators; they are highly distasteful, possessing a coat of irritant hairs, while others bear spines and venom. Those that taste unpleasant obtain their 'flavour' from the plants they eat. The caterpillar of the monarch butterfly feeds on milkweeds. The plant toxin renders both the caterpillar and the adult butterfly particularly repellent to natural enemies.

Venomous caterpillars, such as the North American saddleback, often have warning colours and numerous outgrowths from their body which bear venomous spines. These are either barbed or glass-like, and inject a venom into the predator's skin, usually in the mouth region. If a human picks up one of these caterpillars, the painful sting may cause nausea and an unpleasant swelling at the point of contact.

Of all the insects, bees and wasps are the most commonly encountered 'stingers'. Together with ants, they are collectively known as the

*The wasp's stinging apparatus.*

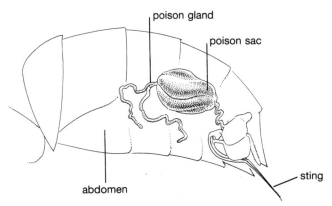

Hymenoptera, and they occur throughout the world; some bumblebees even inhabit the high Arctic region.

Only the females are capable of stinging. In most insects the eggs are laid through a special egg-laying tube, the ovipositor. However, in bees and wasps the ovipositor has been modified in the course of evolution to become the sting. Since males never had an ovipositor, they do not possess a sting, and can be handled quite safely. It is difficult at the best of times, however, to distinguish the sexes, and only of academic interest once stung.

Worker wasps and bees will use their sting to defend their colony's nest from predators. Alighting on the invading animal, the wasp or bee grasps hold with the jaws and thrusts in its barbed sting. The venom gland contracts and the venom shoots through a special canal into the wound. Unlike the wasp, a honey bee's sting is barbed in such a way that it cannot retract once buried in its victim. Once it has stung, the bee flies off, tearing its body away from the sting and stinging apparatus, leaving them embedded in the victim. Later the bee will die as a result. This might appear to be the ultimate sacrifice on behalf of the colony, but all the bees in the colony are genetically related. The workers who defend it are all sisters and all sterile. By giving up their lives to protect their nest mates, they are promoting their own genes through their relatives.

For a human being a wasp or bee sting is always painful, but not necessarily serious. Some people feel it as no more irritating than the sting of a nettle, and have survived multiple stings; others, who are allergic to the venom, may experience unpleasant effects or worse from a single sting. During the years 1949–69 at least 70 people are known to have died in England and Wales from wasp stings, but deaths from bee stings are less frequent.

Several animals are able to withstand such stings and actually feed on various portions of wasps, bees and their nests. In Africa, the honey badger, or ratel, is impervious to the bee's stings, ripping open the nest to consume the honey and bee larvae. It is often led to the nest by a curious bird known as the honeyguide. This bird, too, is indifferent to the angry bees and not only feeds on anything left over by the ratel but also on the wax combs in the nest, having the ability to digest wax by means of bacteria in its gut. In northern Europe, the common badger regularly digs up the nests of wasps and rarely appears to be bothered by their stings. And the predatory honey buzzard, although it does not eat honey, will excavate wasps' nests to feed on the grubs, protected from stings by its dense plumage and scaly legs.

———————————— □ ————————————

The amphibians – frogs, toads and salamanders – contain several poisonous species but no actively venomous ones. All poisonous amphibia use their toxins as a means of self-defence and not to catch prey. The toxins defend them against predators, and most interestingly against the effects of bacteria and fungi.

Terrestrial amphibians respire through their skin and to do this must keep it moist with mucus. The damp surface is an ideal environment for micro-organisms to grow, and protection against them is often provided by special poison-producing glands all over the skin.

It has long been known that amphibian skin contains strong chemicals. Over 3,000 years ago powdered toads were used as ingredients in Chinese and Japanese medicines. The same concoction is still used today for treatment of dropsy, a disease where excess fluid collects in the body tissues. Dogs that bite toads usually come off second best. Within seconds, the toxins from the toad's skin enter the dog's bloodstream through the thin mouth tissues, resulting in retching, foaming at the mouth, and, in extreme cases, death.

Most toads are cryptically coloured, blending with the surroundings, but one group of amphibians, the poison arrow frogs, from the rain forests of Central and South America, advertise their toxicity. These animals are small, only an inch or so in length, and their spectacular coloration makes them look like tiny jewels. The toxins secreted by the skin are among the most poisonous substances known, and these frogs are so named because the Noamana, Choco and Cuna Indians of Colombia use the poison to tip their arrows and blow-darts. The frogs are collected and cooked over a fire until the poisonous, milky skin secretions appear; these are then scraped on to the weapons. The poison is so effective that a single dart can kill a monkey in seconds.

*The poison arrow frogs of Colombia are roasted by South American Indians to extract the poison from their skin. The Indians then anoint the tips of their arrows with the poison.*

Snakes, tortoises, crocodiles and lizards are all reptiles. Together with spiders, snakes are among the animals most feared by man, although most species are harmless. No tortoises or crocodiles are venomous and only two lizards are, namely the gila (pronounced 'heela') monster and the closely related Mexican beaded lizard. Both have a stout, chunky appearance, the gila monster being slightly smaller, reaching 60 cm (24 in) in length, the beaded lizard growing to 75 cm (30 in); and both are inhabitants of the south-western desert regions of North America, including Arizona.

These two lizards use their venom to kill their prey, which comprises small mammals such as mice and birds. They also eat the eggs of birds and other reptiles. The venom glands are modified salivary glands located in the lower jaws. The teeth are specially grooved so that when the lizard bites, the venom runs down the outside of the teeth into the wound. The venom is particularly effective against birds and mammals, but less so against frogs. The prey is located by means of smell, using the tongue and a special gland situated near the roof of the mouth.

Gila monsters and beaded lizards are generally nocturnal and hunt mainly on the ground. Their large, fat tails contain food reserves, utilised when feeding conditions are poor. Although their bite is extremely painful to man, they are not aggressive and most people are bitten while handling them. This can be a frightening experience because when a gila monster bites, it does not let go for some ten to fifteen minutes, during which time it continues to chew, injecting more and more venom into the wound. Symptoms include localised pain and numbness followed by breathing difficulties. Fortunately, however, there are few fatalities. Indeed, man is much more of a threat to these lizards, by keeping them as pets; their existence in the wild is now threatened by collectors.

## WARNING COLORATION

Many poisonous or venomous animals advertise their dangerous defences with bright colours. The poison arrow frogs are one example, while many bees and wasps bear characteristic black and yellow stripes. Most animals quickly learn to keep clear of boldly striped insects, particularly if they sting. In a few species such recognition may be inborn, but in the majority it has to be learned by experience. This poses an intriguing evolutionary problem.

Most predators probably have to sample and kill at least one wasp or bee before they learn to identify it as unpleasant. If a wasp or bee has to die in order to convey the message of its coloration, how did warning colours evolve in the first place? The answer is linked with the fact that many distasteful or venomous animals live in social groups. Such groups consist of closely related individuals, and in evolutionary terms at least it is worth one of them dying to help the others, in order that a predator may learn not to touch them again. Warning colours presumably evolved because such striking patterns are easily recognised and the lesson quickly learned.

Some non-venomous and non-poisonous animals have exploited the fact that predators tend to avoid warningly coloured individuals. These quite palatable species therefore mimic the more dangerous ones and benefit from the similarity. Many flies, especially hoverflies, are banded black and yellow, resembling wasps. And there are several toxic butterflies, like the

*The yellow and black bands of the wasp are recognised as dangerous by man and beast. It is only the females that sting, but it is not easy to tell the difference between the sexes.*

monarch, which have their harmless mimics, such as the viceroy. This form of impersonation, discovered by Henry Bates while collecting insects in South America, is referred to as Batesian mimicry. Other examples include edible cockroaches which mimic ladybirds, squirrels that mimic inedible tree shrews and one grasshopper which mimics the bombardier beetle. This beetle discharges a very hot fluid from a special gland at potential predators. A second form of mimicry, discovered by Fritz Müller, who likewise collected butterflies in Brazil, involves several distasteful species resembling one another. Several species of bees and wasps, as well as butterflies, display Müllerian mimicry.

*Once the gila monster gets its teeth into its prey it will not let go. It chews and bites the venom into its victims, generally small mammals and birds.*

There are about 2,700 species of snake in the world, yet only 50 or so are highly venomous. The most feared of these include the saw-scaled viper, the king cobra, the green and black mambas, Russell's viper, and the lancehead viper or fer-de-lance. Unlike more primitive snakes such as pythons and boas, which constrict and suffocate their prey, these more 'advanced' species subdue or kill their victims by injecting them with venom. Their prey may range from insects, snails and lizards to rodents, birds, birds' eggs and small mammals.

Snake venoms have different effects, some simply weakening or disorienting, others paralysing or killing the prey. But they can also be used against predators. The amazing habit of the spitting cobra of squirting venom for a distance of several feet into the eyes of other animals, including humans, is mainly a defence reaction.

As a rule, snakes have simple teeth, all roughly the same shape, long, pointed and directed backwards when the mouth is closed. The front teeth of venomous snakes may be modified to conduct venom from the glands in the upper jaw to the tip of the tooth. In many species these teeth are hollow along most of their length. For example, in the rattlesnake the upper end of

the tooth near the jaw has a large inlet for venom (the basal orifice), and a smaller outlet close to the tooth's tip (the distal orifice). In some snakes the fangs are hinged and are folded back along the roof of the mouth; when the snake is about to strike, the fangs flip forward into their striking position. The Gaboon viper holds the record for the longest fangs – of the snake's 3 m (10 ft) length, the fangs measure almost 3 cm (1 in).

The snake's venom glands are modified salivary glands, and the venom modified saliva. Snake venom is usually clear yellow and is dangerous only

*The deadly spitting cobra of East Africa.*

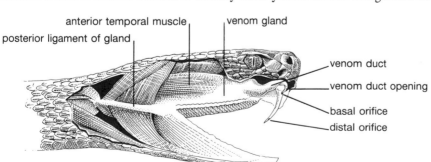

anterior temporal muscle

venom gland

posterior ligament of gland

venom duct

venom duct opening

basal orifice

distal orifice

*A rattlesnake's fangs are neatly folded away when not in use but are swung forward for the strike.*

103

if it gets into the bloodstream. The amount of venom injected varies from species to species, ranging from a few microlitres in highly venomous sea snakes to six or seven millilitres in the diamondback rattlers and pit vipers. In snakes like the black mamba, cobras and Russell's viper, the amount injected in one bite far exceeds that necessary to kill a man. Moreover, snakes never inject all their venom in a single strike, so that they are always ready to bite their prey or attacker again.

Snake venoms work in several ways. Some types of venom damage the walls of blood vessels and also prevent blood from clotting. Others act on the nervous system, blocking nerve transmission and causing paralysis of the heart and lungs. Certain snake venoms contain a mixture of both substances.

Human deaths from snake bites are caused mainly by accident. In the Indian subcontinent, where large numbers of people work on the land, it is estimated that many thousands of people die from snake bites every year. Treatment varies from the most simple cut-and-suck method to the more

*The Gaboon viper of Africa: its large venom glands account for the distorted arrow shape of its head.*

# RATTLESNAKES

The rattlesnakes belong to the viper family which includes some of the most venomous snakes in the world. Rattlers occur in the warmer, drier parts of North America. The body is heavy and fairly thickset, and the contrasting skin surface markings consist of zigzag or diamond-shaped patterns.

The distinctive 'rattle' at the end of the tail is formed from modified scales and increases in size by one rattle section every time the snake moults its skin, which may be up to four times a year. As the rattle gets longer, the end sections start to break off. The maximum number of rattles observed is 20. The tail is vibrated and shaken against rocks when the snake is irritated or alarmed and serves as a warning. The characteristically wide, triangular-shaped head accommodates large venom glands as well as powerful muscles necessary for squeezing the fast-acting venom through a single pair of fangs into the prey's body. Frogs, lizards and small mammals are all hunted on the ground; and the largest rattlesnakes, such as the diamondbacks, which may exceed 2 m (6½ ft) in length, feed on animals as large as a hare.

The eastern diamondback and the western diamondback are regarded as the most dangerous snakes in North America and are responsible for a number of human fatalities every year. As a rule, however, a rattler avoids contact with people and will strike only if cornered. Yet in some parts of the United States, as in Texas, tens of thousands of rattlesnakes are rounded

*The western diamondback rattlesnake is considered to be one of the most deadly snakes in the USA. But this does not justify its ritual slaughter.*

up, killed, skinned and often eaten at barbecues. This is a practice that has threatened several species.

sophisticated use of anti-venins. The latter are increasingly available but specific for each kind of snake, so that immediate and accurate identification of the species is essential. As with bites from other venomous animals, young and elderly people are most at risk; and even if a bite is not fatal, it may cause massive tissue damage, so that hospitalisation is often necessary.

The cobras have marine relatives in the sea snakes, which probably evolved from Australian ancestors. Approximately 80 per cent of the Australian species are venomous. They show a range of survival factors associated with their aquatic life. Some spend their entire time in the water, mating and giving birth to live young in the sea. A few come ashore to lay eggs, usually in caves. Most sea snakes have a compressed, oar-like tail, used for propulsion through the water. When diving, they can also seal their nostrils and mouth. The extreme toxicity of their venom enables them to kill their main prey – eels – very rapidly and thus prevent them from wriggling free.

Fishermen of tropical seas who regularly catch sea snakes display little or no fear of them, handling them with impunity and either killing them or throwing them back in the water. The reason for such nonchalance is that the snakes seldom bite people and, even if they do, rarely inject much venom. Their bites are, however, extremely effective in subduing the prey.

———————————— □ ————————————

There are a few venomous mammals, including certain shrews. Some water shrews and one long-snouted shrew-like animal known as *Solenodon*, which inhabits Hispaniola and Cuba, possess poisonous saliva. Injected when the animal bites, the saliva causes paralysis. Some water shrews feed on relatively large fish or newts, so it is important for them to immobilise their prey rapidly. The *Solenodon* eats insects and the occasional bird, reptile or amphibian. Only one or two young are born at a time and they are transported in an unusual manner by the mother – by hanging on to her greatly elongated teats.

The duck-billed platypus is an even more astonishing animal. And it is venomous, too. When the body of one specimen was first brought from Australia to Britain in 1798, it was considered to be some kind of taxidermist's hoax – a mixture of bird and mammal – which, to some extent, it is. The platypus, along with its cousins the echidnas, is an egg-laying mammal yet suckles its young on milk!

Confined to eastern Australia, the platypus spends virtually its whole life in freshwater streams. It is covered in a dense, waterproof fur and has webbed forefeet and partially webbed hind feet. Its best-known feature is its soft and pliable duck-like beak.

The male platypus possesses on its hind legs a large, reversible spur. This is usually tucked away under a fold of skin but it can be quickly erected and used as a weapon. The hollow spur is connected to a venom gland situated behind the knee, and can inflict an agonising wound. It is employed in defence, and has been known to kill a dog. However, the male platypus also makes use of it when competing for territory along a river, although its effect on rival males is not known.

How the platypus came to acquire this apparatus remains a mystery: few venomous animals use their venom when fighting among themselves, and it has therefore been suggested that the venom gland is a survival factor originally directed against some long-extinct predator. This seems unlikely, however, because one might expect females to be just as much in need of protection as males. It is curious, too, to note that echidnas also possess spurs but their venom gland is non-functional.

The toxic nature of some animals has been known for a long time and man has utilised their venoms and poisons for various purposes. Native peoples in Africa and South America have employed animal toxins to help them catch prey; Kalahari bushmen anoint their hunting arrows with the squashed contents of beetle larvae to produce a slow-acting poison, and the Colombian Indians, as described, dip their blow-darts into the secretion of the poison arrow frog.

*Scientists are now making use of the black widow spider's venom to help them understand disabling human diseases.*

   Yet if many of the toxic chemicals that animals possess have evolved in order to aid survival, either by predation or protection, it may well be that some of these compounds could assist our own survival. Understanding how animal venoms and toxins work may one day help to cure human diseases, and experimentation is under way. Many toxins, such as snake venoms, operate through the victim's nervous system and are being used in the treatment of human nervous disorders. And the venom of the notorious black widow spider, far more potent than that of a rattlesnake, is now being utilised to treat disabling diseases of the nervous and muscular systems. So, given sufficient time, interest and means, it is possible that the hitherto dangerous survival factors possessed by animals may become our own means of survival.

# LIFE ON THE EDGE

Mountain goats live amidst the majestic, windswept heights and steep slopes of the Rocky Mountains of western North America, from Colorado in the south to Alaska in the north. It is a bleak landscape, stretching to altitudes of 4,000 m (13,000 ft) or more, and composed of snow, bare rock, scree slopes and ice fields, with only the occasional bush and clump of grass to relieve the harshness of the scenery. This is an exacting environment for any animal to inhabit. Winter lasts for nine months of the year, when temperatures can fall as low as −50°C, and conditions are made worse by the deadly wind-chill created by high winds. Mountain goats have to contend with narrow ice-covered ledges, deep snow drifts and avalanches. Moreover, their food plants are sparsely distributed, forcing them to take risks in order to get enough to eat. Temperatures during the short summer are somewhat kinder and may on some days reach 20°C. Nonetheless, summer or winter, this is an inhospitable environment and to be able to survive here, the mountain goat needs a special set of structural and behavioural adaptations.

The result is a very distinctive animal, which has only recently been properly classified. Early explorers of the Pacific coast of North America, who coveted the pure white pelts worn by the resident Indians, thought that it was some kind of polar bear. When Europeans saw the animal alive for the first time, they christened it the mountain goat because of its unusual agility on narrow, icy ledges. Scientists did not help matters when they gave it the generic name *Oreamnos*, meaning 'mountain lamb'. The mountain goat is not a lamb nor, despite its common name, is it a goat. In fact, it belongs to a group of animals known as goat-antelopes, or Caprinae. Its

Opposite *Mountain goats in Jasper National Park, Canada.*

109

*The rare Himalayan tahr – a distant relative of the mountain goat – has also adapted to life in harsh mountainous conditions.*

best known relative is the chamois, a native of the European Alps which has also adapted to life at high altitudes on precipitous mountain cliffs. Other less well known relatives are the goral of southern Asia, and the two species of serow: the Japanese serow, which also has a dwarf form on Taiwan; and the mainland serow of the warmer parts of south-east Asia. Fossil goat-antelopes dating back 35 million years closely resemble the serow, and present day goat-antelopes have probably evolved from a serow-like ancestor.

Other, slightly more distant, relatives of the goat-antelopes include the musk ox, the Himalayan tahr, the takin, the ibex and the mouflon or Barbary sheep. All these animals live in harsh mountainous environments, and many of them, like the serow and chamois, are expert climbers. Finally, and still more distantly, the goat-antelopes are related to cattle, true antelopes, gazelles and other members of the Bovid family, which is itself a subdivision of the large and diverse group known as artiodactyls, or even-toed ungulates.

All the even-toed ungulates share certain characteristics, which are important in understanding how the mountain goat has managed to adapt to the particularly inhospitable environment in which it lives. They are all hoofed mammals, with two main toes of roughly equal size; and they are all (with the singular exception of the pig, which is an omnivore) herbivores – they eat only vegetation.

Plant material is poor in nutritive value, and difficult to digest, and herbivores have had to adapt in several ways to make the most of it. The most obvious adaptation is in the structure of their stomachs. These have several compartments – four in the case of the mountain goat. One compartment, the rumen, is a repository for the raw mass of unprocessed leaves, grass stems or twigs that the animal has just eaten. Here, bacteria get to work on it, softening it up and making it ready for chewing. When the period of feeding is over, and the animal is safe from predators, the contents of the rumen, the 'cud', are released back into the mouth, where a set of specialised cheek teeth take over. The cud is thoroughly masticated – the jaw can move from side to side as well as up and down – before it is finally passed into the digestive system where the job of extracting everything of nutritive value will be carried out. As an incidental benefit, the fermentation processes within the stomach generate heat, which is of no small value to an animal living in intensely cold areas.

This ability to live off poor vegetation is likely to have been crucial in the evolution of the mountain goat. Millions of years ago, some serow-like ancestor may have been tempted by the relatively secure niche provided by the steep mountain slopes which are safer from the predators than the plains below. Natural selection would then gradually have produced a hardier group of individuals capable of remaining for longer in the remote mountain regions. Eventually, the mountain goat would have evolved.

The other key characteristic of the even-toed ungulates that would have helped this process along is the structure of the feet. Unlike the hooves of deer, horses and other ungulates (hoofed mammals), which have rigid edges, the hooves of the mountain goat have rubbery, elastic sides. Thus, unlike the others, the mountain goat does not have to walk continuously on the sides of its hooves: when it is leaping or running, the side of the hoof will give, bringing a much greater area of the toe into contact with the

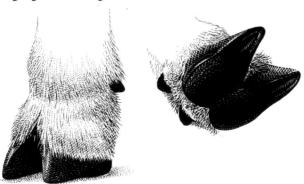

*The very flexible hooves of the mountain goat give it maximum grip on precarious slopes and dizzy ledges.*

111

*The chamois, another mountain specialist, is much lighter and consequently much more agile than the mountain goat.*

ground and greatly increasing the animal's ability to keep a grip on the icy rock. Furthermore, there is a much wider and more flexible gap between the two toes of the mountain goat than there is in other ungulates. Not only does this allow it to 'grasp' the ground more firmly, but it also allows the toes to splay out when the goat is descending an icy slope, giving more effective braking power.

Just how important it is for the mountain goat to be footsure is summed up by Doug Chadwick, an American biologist who lived with goats for several years:

> Take a typical staircase in your home (30–35°), tilt it to match the angle of the sides of Egypt's Great Pyramids at Giza (52°), remove some of the steps, and shovel ice and a couple of feet of snow over the whole affair. Then, as long as this is only a mental exercise, try moving the concocted cliff atop an apartment building. Place it right on the edge so that the next step below the bottom is at least a couple of stories down, because even moderately sloping sections of a mountain goat cliff may have forbidding dropoffs at the base.

Confronted by physical hazards like this, mountain goats are cautious as well as footsure. If you want to see dazzling feats of mountaineering, with animals leaping from ledge to ledge across gaping chasms, then look to the chamois. The mountain goat is built primarily for steady, tireless climbing rather than speed and great agility: its short legs and short body, which allow it to place all four feet close together, also give it great stability on narrow, uneven ledges. Some idea of the mountain goat's steady strength can be gained by comparing its climbing speed with our own. A 500 m (1,750 ft) ascent, which took an experienced human mountain walker 90 minutes, was climbed by a mountain goat in 20 minutes. It is only young, inexperienced animals that engage in spectacular jumps and other daring manoeuvres. If you were to see an adult goat jump like this, it would usually mean that it was in a desperate situation.

The mountain goat's ancestry, as a member of the Bovid family, gave it a good start: it gave the animal the means of subsisting on the poor mountain vegetation, and it provided it with feet that could be adapted to moving around in the precipitous terrain. But much more is required before evolution can produce an animal so finely integrated with its environment. Many other factors – anatomical, behavioural and social – play a part in the continuing survival of the mountain goat.

*A bighorn lamb tries its mountaineering skills. The mortality rate for young mountain mammals is very high. Many fall to their deaths.*

Keeping warm is obviously a prime necessity in the freezing temperatures of the Rocky Mountains, and the coat of the mountain goat, as the Indians and European explorers discovered, fulfils this need superbly well. Only the musk ox has a denser covering of hair. The mountain goat has a long, shaggy coat, brilliant white immediately after the autumn moult, but dirty yellow by the end of the winter. The thick mane and the 'pantaloons' on its legs, together with its small, sharply pointed black horns, give the goat its characteristic and unique appearance.

The coat consists of an outer layer of guard hairs, up to 20 cm (8 in) long, overlying a thick layer of fine, cashmere-like wool. The guard hairs are hollow, trapping air within and between them, and the underfur also traps air. Air is one of the best of all insulators against heat loss when not allowed to circulate, so, between them, these two layers provide a high degree of protection from the biting cold. What might seem surprising is that the coat is white. White reflects the heat of the sun rather than absorbing it, whereas black does the opposite. So, except during a few warm days in summer, a white coat would seem to put its wearer at a disadvantage. For smaller animals like the mountain hare, the benefits of camouflage against a background of snow no doubt outweigh the disadvantages. But the mountain goat is big enough to have relatively few predators. The answer to this puzzle may have been provided by some research carried out by American engineers on another large white mammal of very cold regions, the polar bear.

The first thing the researchers discovered was that special infrared cameras, designed to pick up the heat given off by warm-blooded animals, failed to detect polar bears – the insulation provided by their coats is too efficient. Much more interesting, however, were the results of further research into the hairs themselves. The bear's hairs, like those of the mountain goat, are hollow. They are also transparent to ultraviolet light which they conduct, in the manner of an optical fibre, from the tip of the hair to the skin, where its energy is converted into heat. The hairs are transparent only to ultraviolet light: other light in the visible range is reflected, which is why the coat appears white. This remarkable piece of natural technology is now being used in the design of more efficient solar panels.

The coat of the mountain goat has some other interesting properties. Hunters often complain about how difficult it is to get a dead mountain goat down from a mountain – it simply will not slide easily across snow. A closer look at the guard hairs, under a high-powered microscope, reveals why. The hairs have overlapping scales, which gives them a rough texture and increases the friction between goat and snow. If you can imagine being on a steep, snow-covered slope, below which there is a 1000 m (3,500 ft) sheer drop, would you prefer to be wearing a shiny nylon cagoule or a rough old woollen jumper? The long, rough guard hairs may have saved many a mountain goat from the consequences of an accidental slip.

*The thick shaggy coat of the mountain goat prepares it for the rigours of the Rocky Mountains' winter.*

A further response to cold is shown by the variation in body size from the north to the south of the mountain goat's range. There is little difference between males and females, but whereas the largest males in the colder, northerly parts of the range may stand 1.2 m (4 ft) at the shoulders and weigh 140 kg (308 lb), those from the warmer south reach a maximum of only 70 kg (154 lb). The reason is simple: the larger the animal, the smaller its surface area in relation to its volume, and so the smaller the relative area through which heat can be lost. Thus, despite the difficulty of feeding a larger body, the survival value of size becomes greater as it gets colder. Many other animals show similar geographical variation in size.

*The Dall sheep, another goat-antelope, can be found from British Columbia to Alaska in North America.*

The annual cycle of the mountain goat, as well as its social life, is determined to a considerable extent by the need to find food. Living where they do, mountain goats cannot afford to be fussy about what they eat. Depending on the nature of the local vegetation, they may subsist almost entirely on grasses, or on the leafy branches of Douglas fir. They also eat chokecherry and serviceberry leaves, fern rhizomes, the lichens hanging from branches, and mountain ash twigs and leaves. In their impoverished mountain habitat, they cannot afford to waste anything. To pick and choose which plants they were going to eat would put them at a disadvantage since they would waste so much time looking for food, rather than eating it.

The poor quality of most of the mountain goat's food, together with the specialised digestive system required to cope with it, imposes a strict daily routine on the animals, which varies little throughout the year. A bout of feeding early in the morning is followed by a lie down during which they

## HUNTING MOUNTAIN GOATS

Deer, antelope, buffalo and other grazing mammals have long been hunted by man for their meat and hides and for the other products they yield. In earlier times in North America these animals provided the Indians with many of their daily needs. But it was commercial hunting by European settlers that devastated the animals' populations. The classic case is that of the American bison, reduced from millions to a handful of individuals, and saved from extinction by a hair's breadth. Men continue to hunt in North America, rarely these days for subsistence, but more usually for sport. But the effects can still be devastating. The mountain goat population was reduced from an estimated 100,000 to 50,000 between the years 1961–1981.

The mountain goat hunting season lasts 11 weeks from 15 September to the end of November. Hunters must obtain a permit and there are bag limits on the numbers of goats that can be shot. Despite these restrictions and regulations, goat numbers have continued to decline.

There is a good reason for this. Whereas other ungulates like white-tailed deer and moose produce a surplus of young each year, allowing a certain proportion of the population to be harvested without affecting the breeding stock, the same is not true of mountain goats. Compared with deer, mountain goats have a slow reproductive rate. Whereas three-quarters of all female deer produce young each year, only a little over half of all female goats do so.

Moreover, while twins are common among deer, they are virtually unknown among goats. Add to this the fact that among mountain goats most of the winter mortality falls on the youngest animals and it is not hard to see why there are few surplus goats for the hunter's guns.

Several other factors have contributed to the mountain goat's decline. Firstly, their open habitat makes them extremely visible to human hunters. They can be seen from several miles away, and they can also be shot (not necessarily killed) over considerable distances with high-powered rifles. Secondly, the hunting season also coincides with the rut. As we have seen, this is a crucial part of the mountain goat's annual cycle and one that is easily disrupted by hunting. Hunting statistics can be misleading since they record only those animals actually taken by hunters. There are no estimates for the wounding rates of mountain goats, but for other game species it may be as high as 40 per cent. Among the hunting fraternity the mountain goat has a reputation for being difficult to kill cleanly and is sadly renowned for its ability to 'carry a lot of lead'.

Because of the competition between males, relatively few of them ever breed, whereas almost all females do. Hunting would therefore have less of an impact on goat populations if hunters avoided killing females. Unfortunately, however, the differences in appearance and behaviour between male and female goats are so subtle that some hunters are incapable of distinguishing between them.

chew the cud. This is followed by a midday feed and another rest, which is in turn followed by a prolonged evening feed. This pattern of alternate feeding and resting is characteristic of all grazing animals. The daily routine is disrupted only by extreme weather and by the rut, and even during the rut it is only the males whose activity is much changed.

Even though the mountain goat can eat more or less any plant material, there is seldom enough to go round, and the availability and whereabouts of food vary throughout the year. The goat's movements and social interactions show a similar seasonal variation.

Food is at its most plentiful during the short summer months, when the snow melts to uncover the higher mountain pastures. At this time of the year, the goats can forage fairly freely and will cover relatively large areas. During a typical summer's day, a mountain goat may travel around a kilometre in search of food. After this brief period of comparative plenty, the snow returns, covering the high pastures in a blanket too thick and hard for the animals to scrape away with their feet. They now have to resort to narrow and exposed ledges, and the shoulders of the mountain, where the continuous high winds prevent a thick cover of snow from settling. Foraging here is a little easier, but it is also more hazardous. The animals' daily movements become much more restricted – covering an area of no more than a hundred metres square. They become familiar with the area, using regular paths and tracks, so minimising the constant risk of falling. Finally, as the worst of the winter sets in, the goats move down to the more wooded foothills, where the trees moderate both the wind and the snow. Here, they are still able to dig through the snow to uncover whatever sparse winter vegetation remains below.

For most of the year, males and females live and feed apart. The separation is at its most apparent in the summer, when the males feed on rougher pastures higher up the mountain. They are often solitary. In contrast, the females tend to live in small groups, with their offspring, foraging on the better pastures lower down. In general, the areas occupied by solitary mature males are less rich in food than the female ranges and this may be a deliberate strategy on the male's part. The theory is that if he stayed with the females throughout the year, he would compete with them for food and would probably take the best of what was available. By leading a solitary existence, he avoids competing with the females that he has fertilised. In other words, by staying away he gives his offspring a better chance of survival.

Compared with many other grazing mammals, like the North American bison, African wildebeest and many other antelopes, which all live in large herds, mountain goats are relatively unsocial. The groups formed by the females and kids are small, and solitary animals, usually males, are far from uncommon. Females with very young kids also tend to keep away from the herd. Obviously, from time to time, several groups will meet at a good feeding site or at a salt lick and larger gatherings occur, but small groups are the general rule.

There are both advantages and disadvantages in group living. The chief advantage is that it offers greater protection against predators. The disadvantage is that a large number of animals feeding in a small area will exhaust the food supply much more rapidly than would a small number.

Unlike many carnivores, for whom a single meal a day is often sufficient, herbivores have to spend much of their time eating. While feeding, they are vulnerable to predators and have to stay constantly alert: head down to feed, head up to look, head down to feed, and so on. For animals in a group, each individual spends less time being vigilant and more time feeding, but the overall level of watchfulness is increased simply because there are more pairs of eyes looking out. Mountain goats may be relatively safe from predators, but they are not totally free of them and the kids are particularly vulnerable. Large billy goats alone may be quite capable of

*With the onset of summer, the snow starts to melt and the mountain goat begins to lose its thick, shaggy winter coat.*

looking after themselves. Thus, females and kids clearly derive some benefit from living in groups. The reason that the groups are not larger is to do with the sparsity of the food supply. The mountain goat's food is well spread out, and nowhere abundant; larger groups would eat more and force the goats to forage over a wider area. They would spend a disproportionate amount of time moving between feeding areas and, in the bleak heights of the Rocky Mountains, there is little time to spare.

Mountain goats have to keep alert, for although the mountains have less predators, there are still a few potentially dangerous ones about. These include the grizzly bear, cougar, wolverine, wolf, coyote, and bald and golden eagles. All of these species have been recorded eating mountain goats at one time or another, but whether they are regular or serious predators on the goat is another matter. Doug Chadwick saw grizzlies chase mountain goats on a number of occasions but never saw one caught.

In fact it seems likely that mountain goats are very successful at avoiding all these predators. If a bear suddenly appears, the goats head for precipitous ledges where the less agile predator cannot follow. If they are cornered by a predator, mountain goats will not hesitate to use their horns to defend themselves. The goat's horns are extremely sharp and potentially lethal, and there is more than one record of a grizzly bear being fatally wounded by a goat. There was also a case in which a hunter's pack of hounds, initially on the scent of a cougar, ended up chasing a large male goat. The goat was cornered and made a stand against the dogs. By the time the hunter caught up with them, the dogs had all been skewered by the goat.

*Young kids are sometimes seized by bald eagles.*

The predators that mountain goats appear to fear most are eagles. Goats will even toss their horns at the shadow of an eagle flying overhead, and will chase eagles off ledges. Eagles are primarily interested in very young goats and there are several observations of golden eagles and bald eagles snatching young kids from beside their mothers. As well as taking kids, eagles also try to flush or knock older goats off ledges. No one has seen a mountain goat killed in this way, but in Europe, golden eagles regularly knock chamois from cliff ledges in order to feed on their carcasses. On one occasion a biologist found a two-year-old goat and a golden eagle locked together at the foot of a cliff. The eagle still had its talons in the goat, and the goat had its horns embedded in the eagle.

While bears, wolves and the like undoubtedly eat mountain goats, they probably do so through scavenging for carcasses rather than through direct predation. The most serious enemy of the mountain goat is snow, or more precisely avalanches. In late winter, avalanches are common and many goats are swept away. Predators like bears, coyotes and lynx regularly patrol the bottom of avalanches looking for carcasses.

Snow kills goats in other ways too. Foraging for food in the depths of winter often means travelling through deep snow, and this takes its toll of the goat population. Small, immature goats suffer most; they find moving around in deep snow most difficult and demanding. Their small size means that they lose heat more rapidly than the adults. Moreover, since immature goats are subordinate to older ones, they may well have to give up any patch of food they find to a larger, more dominant animal. All in all, the young have the odds stacked against them, especially in a hard winter. Normally, about half the kids born each year will be dead by the following spring. In a severe winter, the figure can reach 80 per cent. Although this may seem harsh it is actually natural selection in action. Hundreds of generations of goats have been subjected to these ordeals each winter, and the best adapted animals have survived.

Severe winters take their toll of adults too. Females that have struggled to obtain enough food throughout the winter and emerge in spring in poor condition, may fail to conceive, or may give birth to a weak and sickly kid. Snow, however, is not always a disadvantage to mountain goats. It can help camouflage them and, paradoxically, it can sometimes help them stay warm. Snow is a good insulator, and the temperature inside a snow cave will remain at just above freezing, regardless of the temperature outside. When conditions are bad, mountain goats will dig a deep bed in the snow, or hide in caves and under rocks.

*Grizzly bears feed on salmon moving up river to spawn. When times are hard a grizzly may take a mountain goat.*

In these particularly harsh circumstances, it seems likely that the mountain goat has adapted to living in small groups as a compromise. It is the solution that offers the best balance between advantages and disadvantages – the animals gain the greatest degree of mutual protection from predators whilst sacrificing the least in terms of their ability to find food in the inhospitable terrain.

———————————— □ ————————————

Wherever animals live in groups, some kind of hierarchy is established. Within the mountain goat community, this leads to continual tension. Mountain goats are unusual in that mild displays of aggression are especially frequent. They continually threaten, nudge and jostle each other; what they are doing is competing for rank. The social system is based on a hierarchy within which the most dominant individuals take the greatest share of food. The largest individuals, usually the oldest ones, tend to be the most dominant, and the smallest and youngest the least. Males usually dominate females, but adult males are absent from the herd for most of the year. The quest for dominance starts at an early age, and even kids spar with one another. As a young animal grows its rank changes, but not automatically; it must fight for every increase in dominance it achieves. Females also compete with each other for rank, on their own behalf and on behalf of their kids. Dominance relationships thus exist both between and within age classes. Dominance can make the difference between life and death, or breeding and not breeding. It is in each individual's interest to defend its place in the 'peck order' vigorously, as well as to challenge those higher up – hence the almost continual challenging that goes on in goat society.

Actual fighting between individuals is rare. The damage that goats can inflict on each other with their sharp horns is so great that aggression has become highly ritualised. Two goats will circle head to tail in a display of strength, tossing their heads and posturing and grunting at each other. They seldom come to blows. In one study, less than 5 per cent of the aggressive encounters involved the goats touching each other. After an aggressive incident, the subordinate goat will retreat in a low-rumped posture – trying to avoid a blow to its flanks. In an extreme case, where a subordinate animal is cornered by another, it will actually squat – the ultimate sign of submission in goat society.

When fighting does occur – as when two evenly matched individuals meet – it seldom results in injury. Unlike true goats, mountain sheep and musk ox, mountain goats rarely direct their blows to their rivals' heads. In fact, one of the features that sets the goat-antelopes apart from their relatives is the relatively thin and fragile skull. In contrast, such animals as the musk ox, ibex, Barbary and bighorn sheep, possess large horns attached to a massively reinforced skull. In these species, head-butting is the normal form of fighting. Mountain goats, on the other hand, aim for the

# HORNS

Mountain goats possess slim, dagger-like horns. Some of their cousins, however, bear monstrously large, curved horns, designed for butting rather than jabbing. The fights of mountain sheep both look, and sound, spectacular as males clash horns with a resounding 'crack' that echoes through the mountains. Fighting occurs during the rut as males compete for dominance. As in mountain goats and other animals, fighting occurs only between similarly matched individuals. If two mountain sheep of markedly different size meet each other, the smaller one will behave as a subordinate and retreat. If, on the other hand, the two sheep are of similar size they can assess each other only by fighting.

The sheep often use gravity to increase the force of their attack: a male that stands above another on a hillside is at a distinct advantage; he can drop on to his opponent using the full weight of his body. The huge horns are used both as a weapon and as a shield. A sheep that is attacked will spin round to make sure that the attacker's horns clash with his own. The thick, air-filled bone of the skull helps to cushion the blow. This dual function of horns is common to almost all ungulates, and individuals with a damaged or lost horn can both inflict and receive severe injuries.

The diversity in the size and shape of ungulate horns and antlers is remarkable. The most extreme form was the now extinct Irish elk, with its 30 kg (66 lb) pair of antlers. It is not clear why horns and antlers should differ so much in shape, but we do know something about why size varies so markedly. Among those animals where a few males mate with most of the females and where, as a result, competition between males for females is intense, weapons such as antlers, horns and teeth are particularly well developed. Most primates, for example, use their canine teeth for fighting. In those species like the baboons, where males monopolise a group of females, the male's teeth are much larger than those of more monogamous species. In the same way, the most polygamous species of goats and sheep have the largest horns or antlers. Although the Irish elk is no longer with us, just from the size of its horns we can be fairly certain that it was polygamous, and that a few males mated with most of the females.

Below left *A goat's age can be determined by analysing the growth rings around the horns.*

Below right *A male bighorn sheep with a particularly fine set of horns.*

opponent's rump and, just as the musk ox's skull is enormously reinforced, so too is the skin on the mountain goat's rump. It is about 21 mm (nearly an inch) thick, compared with about 3 mm (⅛ in) over the rest of its body. (As a comparison, the rump skin of a domesticated goat is about 2 mm (¹⁄₁₂ in thick.) As a result, the goat is well protected against its opponent's horns, and a fight will usually end with the subordinate animal signalling submission, before any real damage is done.

Because aggression is ritualised in this way, it poses few threats to the individuals involved. Most aggressive encounters serve more to confirm the existing social hierarchy than to change it. However, there is one time of the year when the stakes are much higher and the rules are correspondingly different. This is during the rut. Fighting, which is primarily between mature males, takes on a much more serious tone. The sole purpose of a billy goat's life is to sire offspring, and the more he fathers, the greater is his success in evolutionary terms. A goat that produces no kids is an evolutionary failure. Compared to mere jostling for social position, fighting for females is a very important matter indeed.

*The ibex is a distant relative of the mountain goat. The sound of their horns clashing during the rut can be heard up to 8 km away.*

The rut starts in late October, after the mountain goats have completed their moult, and are sporting their new winter coats. A solitary males will patrol the cliffs in search of bands of sexually mature females. Once he has found a group, the male stays with it but at a reasonable distance of perhaps 30 or 40 m (100–130 ft). The male's usual pattern is to position himself on higher ground, above the females, keeping careful watch on their every move. He may shadow them for two weeks, gradually getting closer and closer. During this time he does not feed, but relies on his body fat, accumulated during the period of good summer feeding – this prolonged fast may later prove very costly.

Every so often the male will dig a rutting pit in the earth and wallow in it, spreading his goaty odour around. He also rubs his scent glands, located just behind each horn, on to clumps of grass and bushes. His painstakingly slow approach is necessary because he might otherwise frighten off the smaller females. Eventually, after two or three weeks, he has moved so close to the females that he virtually touches them.

*A male bighorn sheep curls back his upper lip in an attempt to determine the reproductive state of a female close by.*

Up to this point, the females have assiduously ignored the male, but soon he starts testing them to see if they are ready to mate. He sniffs and licks their genital areas and then curls back his upper lip in a characteristic facial gesture, known as *flehmen*. The purpose of this lip curl is to lodge the scent or taste of the female's urine in a particularly sensitive part of his nasal cavity, the Jacobson's organ, so that he can assess the female's reproductive condition. If the female is 'on heat' – in oestrus – she will accept the male and mating takes place. If she is not ready, she will move away from the male and the male will not attempt to mate with her. As in many mammals, oestrus lasts a short time; just 48 hours in the mountain goat. Since all the females in a group tend to come into oestrus at around the same time, the male goat will have a busy two or three days.

Given the amount of time that the male has invested in his group of females, it is hardly surprising that he is ready to defend them vigorously against other males. If another male approaches, it is usually chased off immediately. However, if the intruder is of a similar size or larger, then he might challenge the 'owner'. At this stage aggression is still highly ritualised, but if one animal fails to submit then serious fighting can occur. The goats lunge at each other with their horns and try to flip each other off the ledge. If the weaker goat gets itself into a position from which it cannot escape, it may be gored, and indeed may be killed. Valerius Geist, a well-

125

respected mountain sheep biologist, once found a mountain goat that had been killed by another and had no less than 32 horn punctures, including several injuries to the lungs and heart.

Although the rewards of breeding may be high for males that successfully mate with several females, the costs may be substantial too. The long fast during the rut, together with fighting and mating, may leave the male in poor condition for the winter. There are invariably more mature female goats around than males, and the inference is that the males, as a result of their arduous rut, are more likely to die during the winter. Once again, natural selection is at work, for only the strongest males will survive.

After the rut the males separate from the groups of females and begin their more or less solitary existence once again. The kids are born in the next year with the coming of spring and summer, so that they can take full advantage of the brief period when food is relatively plentiful. Only a single

## MATING SYSTEMS

The extent to which males can monopolise females for mating depends upon how females are distributed. If females are well spaced out then a male would have great difficulty in monopolising more than one at time. If females are concentrated together then it is much easier for a single male to herd them and defend them from other males. When this happens competition between males for the right to breed is intense and results in a marked difference between the sizes of males and females. At the other extreme, where males generally mate with only a single female, the sexes are much more similar in size and appearance.

What determines the spacing of females? The answer for most ungulates appears to be their food supply. The mountain goat's food supply is thinly distributed over the mountain slopes and cliff ledges, and so are the goats. As a consequence they typically occur in very small groups and males therefore mate with fewer females each year. Competition between male mountain goats is rather less than in some other species and as an evolutionary result the sexes are very similar in size and appearance.

Amongst the deer, these patterns are well defined. The muntjac and roe deer are browsers, living either singly or in very small groups throughout the year in dense woodland. At the other extreme are those species like sika, Père David's deer and wapiti, which eat mainly grass and live in open environments. These deer tend

*The male roe deer's branched antlers are used in the breeding season when rutting against other males. Although the female may be mated in July she will not give birth until the following May after a gestation period of almost 300 days.*

to form large herds, and during the breeding season males defend large harems of females. As a result, males in these species are usually noticeably larger than the females. This difference in size between the sexes has evolved because males have to fight in order to obtain and maintain a harem, and bigger males are better fighters.

*A female mountain goat licks minerals from the soil. The mother makes sure she is always downhill from her kid in case it falls.*

kid is born at a time; twins are extremely rare. Slightly wobbly at first, the kid soon learns to stand and to stay close to its mother. In fact, the young goat will remain beside its mother for at least a year. The mother will protect it from predators, such as eagles, and will try to ensure that it doesn't fall down the mountainside by always standing on the downhill side of it. It is more likely that a kid will die than reach sexual maturity at three years old. Nevertheless, the mountain goat maintains a tenuous but firm hold in its cruel environment, even though catastrophe is never far away. For young and adults alike, it is, both literally and metaphorically, a life on the edge.

127

# WHITE WATER, BLUE DUCK

There are over a hundred different species of duck known to man throughout the world. One of the most common of all of these is the mallard. Found throughout the northern hemisphere, it is the archetypal duck from which most domestic forms have arisen. This most adaptable of species belongs to the group of ducks known as dabbling ducks, so known because they either dabble or comically 'up-end' for their food. But, by taking a look at the terms used to describe other types – freckled ducks, steamer ducks, shelducks, eider ducks, sea ducks, stiff-tailed ducks, diving ducks, perching ducks – one can see that ducks occur in all sorts of shapes and sizes and occupy a wide variety of habitats.

Although not a separate group in itself or even a distinct tribe, there are a few ducks that have adapted to life in the fast-flowing, turbulent, white waters of streams and rivers. They include the blue duck of New Zealand, Salvadori's duck from Papua New Guinea, the torrent duck of South America, and the African black duck, which are related to dabbling ducks, and finally the harlequin duck, which is a diving duck. Their similar lifestyle has resulted in a number of shared characteristics which have evolved in response to the testing environments in which they live. This chapter will look at their remarkable combination of 'white water' survival factors in greater detail.

*The fast-flowing Manganui A-Te-Ao river in North Island, New Zealand, is typical blue duck habitat.*

The blue duck inhabits both the main islands of New Zealand, occurring in the central parts of North Island and western parts of South Island, both of which have remote mountainous areas intersected by tumbling white-water streams. The plumage is slate-blue, with a patch of beautiful chestnut feathers on the breast. Unusually for ducks, both sexes look much alike, although the male is slightly larger. The legs and feet are dark brown and the eyes are brilliant yellow; and whereas the eyes of most ducks are positioned on the sides of the head, the blue duck's eyes are directed

129

forwards. This gives the bird binocular vision, a survival factor in that it is better able to pinpoint and capture the tiny, fast-moving invertebrates which are the principal item of its diet.

Another important and striking survival factor is the blue duck's highly distinctive beak. It is mainly white, with a pair of fleshy black flaps forming a fringe around the tip of the upper mandible. The duck forages by pushing its beak under stones and into gravel in search of its food. Apparently the flaps cushion the edges of the beak while feeding, and they may also play some part in sensing where the prey is situated.

A third survival factor is the presence of a prominent bony spur halfway along each wing. Several other birds, such as Salvadori's duck, the mute swan, the spur-winged goose, the spur-winged plover, the black-necked screamer and the jacana, are also equipped with similar spurs, which are invariably used for territorial fighting. What they do for the blue duck, however, is not entirely clear. They may be used in territorial fights, but it is far more likely that they simply help the ducks to climb over and between slippery rocks.

*The blue duck's specialised bill enables it to search for food under rocks and stones. The fleshy lobes at the front cushion the bill against damage.*

The blue duck was at one time well distributed on both islands but its present range is much more restricted and its population, numbered in hundreds rather than thousands, is confined to only a few rivers. The native people of New Zealand, the Maoris, who named the duck 'whio', because of its call, hunted it widely for food; the women also wore the colourful breast feathers around their necks as decoration. The duck was virtually a sitting target, too, for the guns of the early European colonists of the 18th and 19th centuries, forming a part of their diet. What is more, there is evidence that the duck was eaten by even earlier settlers of New Zealand, the Polynesians, who hunted moas, the flightless, ostrich-like birds that are now extinct. Charred remains of blue ducks have been found in ancient middens in areas of Polynesia which are outside the species' present range.

The pioneers from Europe brought not only guns to New Zealand but also a host of familiar birds and mammals from 'home', never imagining how much damage these would do to the country's indigenous fauna.

Before the arrival of the Europeans there was only one mammal – a bat – in New Zealand, and certainly no ground-living predators. But among the animals introduced by the settlers were rats, pigs, dogs, polecats, stoats and weasels, which hunted native ground-nesting birds such as the kakapo, a flightless parrot, and the blue duck. Additionally these birds were threatened as their forest and woodland habitats were progressively cleared to make way for farmland. The felling of riverside trees allowed sunlight through which warmed the waters, but it also increased the rate of erosion along the river banks. This may well have reduced the numbers of aquatic insect larvae upon which the ducks normally feed. The overall result, in any event, was a sharp decline in the blue duck population.

Numbers have further decreased as a result of the more recent damming of rivers for hydroelectric power; and even the seemingly innocuous introduction of trout may have proved harmful. Trout are ferocious hunters of water invertebrates such as mayfly larvae, and possibly compete with the duck for food. Whether or not the competition is fierce, it is surely

*A pair of blue ducks with their four young chicks. Male and female blue ducks probably mate for life and rear their offspring jointly.*
*Two pairs of eyes are better than one when it comes to watching over young ducklings in a torrent.*

no coincidence that blue duck are nowadays more plentiful in rivers that are free of trout.

For all these reasons, it seemed likely at one stage that the blue duck would soon join the ranks of the 50 or so bird species in New Zealand which, like the moa, have become extinct. The situation is still precarious, but provided their white-water river habitats are preserved intact, future prospects are reasonably hopeful.

Food for the blue duck consists principally of the larvae of aquatic insects such as mayflies, caddisflies and stoneflies – creatures that are common everywhere in clean, highly oxygenated, swift-moving mountain streams. During the evening the larvae let go of the rocks and boulders they have been clinging to all day, and move downstream with the current. The opportunistic blue ducks therefore do most of their feeding at this time, dabbling in the shallows with head and neck submerged to catch their prey. In slightly deeper water they will up-end, like the mallard, and grub around for the larvae; and where the water is deeper still, they may dive and vanish from view completely for up to 20 seconds at a time, swimming underwater by paddling with their webbed feet and half-opened wings.

*Adult blue ducks and their large chick. Blue ducks have a bony spur halfway along each wing which is thought to help them to clamber on to and between slippery rocks like these.*

## MATING STRATEGIES IN DUCKS

*Most ducks, like the mallard, show a marked difference between the sexes with the male more brightly coloured than the female.*

The male and female of most familiar ducks display a marked difference in plumage. The male mallard, for example, is brightly coloured while the female is plain brown. The explanation is that the drake plays virtually no part in breeding, other than merely mating, whereas the incubating female is more vulnerable to ground predators. So in response to this heightened risk she has evolved sober camouflaged plumage.

Because of the higher female death rate there is fierce competition among males for a mate. Each individual must spend a considerable amount of energy attracting and then keeping a partner. He will start his search in the autumn, after the moult; and once paired, he has to stay close and protect her from the advances of rival males. He is particularly attentive in the ten days or so before egg laying, for this is when she can be fertilised. His constant presence enables her to feed in peace and allows her to obtain enough nutrition for the eggs to form properly. Yet once she has laid her clutch, the male loses all interest in her and gives her no assistance in tending the eggs or rearing the ducklings. This behaviour is quite at variance with that of the white-water river ducks.

The reason for such apparent negligence is that because mallards breed in relatively calm and benign surroundings, the female is quite capable of caring for the young herself. But in

the turbulent and hostile environment occupied by river ducks, both parents are obliged to care for the young until the latter are completely independent.

Blue ducks differ markedly from most other ducks in their breeding behaviour. Males and females probably mate for life, rear their young jointly and live together on their territory all year around. Few other ducks defend territories throughout the year. Blue ducks patrol stretches of river that measure approximately one kilometre in length. These stretches always contain rapids, where food is most abundant, and are separated from neighbouring territories by areas of still water. Males emit their distinctive whistling calls when establishing or maintaining boundaries, usually at dawn and dusk, and sometimes fly up and down to remind neighbours of their presence. Any intruding birds on the river tend to remain inconspicuous at such times and if detected are rapidly evicted, as a rule.

Breeding may take place in almost any month of the year but most clutches of eggs are laid in September – the southern hemisphere's spring. Before the eggs are laid both sexes perform a variety of courtship displays,

*This particular blue duck nest with its clutch of five eggs has just been flooded, a common hazard in late spring.*

and mate repeatedly. They swim around in unison and call to each other, the male frequently nuzzling the female with his bill, which has now turned from ivory-white to pink.

Experts who have made a detailed study of the courtship ceremonies of the commoner duck species see the similarities and differences as possible clues to how ducks have evolved and how they are related. The displays of the blue duck indicate that it is very different from most other ducks but suggest that its closest relatives are the dabbling ducks like the mallard. Even so, there is not yet enough evidence to say exactly where it fits into the family tree.

Once the birds are in breeding condition, the female looks for a suitable nest site, which may be the same one as was used the previous year. As a rule the nest is located some distance from the rapids, either in a cave or burrow along the river bank or in the open, concealed beneath vegetation. It is composed of dead plant material and is sparsely lined with the rather coarse down that the female plucks from her own breast.

A single clutch of about five eggs is produced each year, unusually small for a duck. The mallard, for example, lays 10–11 eggs. Those of the blue duck, however, are a good deal bigger, and this probably constitutes an important survival factor. A large egg provides room for the embryos to develop large, sturdy feet. The ducklings will need these, for within hours of hatching they must be able to swim in the cascading waters that surround them. A large egg also takes longer to incubate – 32 days compared with the mallard's 27 days. Because of this prolonged incubation there is a greater risk of attack by a ground predator such as a stoat. Although the male stands guard near by, the female blue duck incubates the eggs and is especially vulnerable. This may explain why there are normally more males than females in a typical blue duck population.

Predators do not represent the only threat to the brood. Storms are frequent in late spring and flash floods can transform the rivers into raging torrents, overflowing the banks and flooding the nests. Although the level soon subsides, the sheer force of the current may sweep the entire clutch away. If this happens the adults will not breed until the following year.

All being well, however, the duckings hatch simultaneously. They are boldly marked with dark grey and white down, and their sex is evident from the coloured spot near the base of the tail; males have a chestnut spot, females a fawn one. This is yet another singular characteristic of the blue

*Two different egg-laying strategies. Above, the mallard lays 10–11 small eggs. Below, the blue duck lays fewer, but much larger eggs. Though the blue duck produces fewer chicks per clutch they will be better developed when they hatch. This will give them a better chance of survival in the fast-flowing rivers.*

135

Opposite *A wildlife ranger from the New Zealand wildlife service prepares to release two adult blue ducks. They have been ringed in order to discover more about their range and behaviour.*

*The blue duck incubates its eggs for a relatively long period. The large eggs take almost five weeks to hatch.*

duck, for there are very few bird species in which the sex of the young can be determined at such an early age.

On the river, the parents, vigilant against possible dangers, never let the young stray far from their side, guiding them to the best feeding grounds. Occasionally the mother will carry them around on her back. This behaviour is unusual and occurs in few other water birds, notably the mute swan, the great crested grebe, the goosander, and, it is thought, in the Salvadori's duck.

Although the ducklings can already catch their own food, they grow fairly slowly (perhaps because food supplies are restricted), do not acquire their full juvenile feathering until they are two months old and are unable to fly until several weeks after that. At this stage their beaks are dull grey and their eyes brown, rather than yellow. It is five or six months before they begin to resemble the adults.

The young birds are reluctant to leave their parents' territory. Murray Williams of the Department of Conservation, Wellington, and Jan Eldridge of the University of Minnesota have both studied the question of how long they stay and where they eventually go. On the Manganui A-Te-Ao river of North Island, marking of individual adults and juveniles has established that the young remain with the parents for 90–100 days, but what has not yet been determined is where they go when fledged and how they

eventually find mates. It is already clear, however, that the ducklings' mortality rate is high. If they are not swept away by the strong current, they may be eaten by river eels or killed by birds of prey. It has been verified that of each brood, on average only two ducklings will survive to adulthood.

*The back of this great crested grebe acts as a floating nest for her zebra-striped offspring – to protect them in the early days of life. This unusual behaviour also occurs in the blue duck and, it is thought, the Salvadori's duck.*

Even more of a mystery than the blue duck is the Salvadori's duck of Papua New Guinea. It was named in honour of Count Tomasso Salvadori, an Italian biologist whose special interest was the birds, particularly the ducks, of Papua New Guinea. Although the duck was discovered in 1894, the first nest was found only in 1959. Study of this species is particularly difficult as it lives in rushing streams in the most remote and inaccessible regions of Papua New Guinea, at altitudes of 500–4,000 m (1,640–13,000 ft).

Salvadori's duck is smaller than the blue duck, closer to the size of a mallard. The sexes are similar in plumage and dimensions, with a slim body and a pointed tail. Like the blue duck, it has forward-looking eyes, affording the binocular vision so important for catching small, agile invertebrates. The duck dabbles and probes in fast-flowing eddies for its prey, sometimes diving and staying underwater for about 12 seconds.

As with the New Zealand blue duck, duck and drake remain together throughout the year; and there is a long breeding season, eggs being laid at any time from May to September, which is winter in the southern hemisphere. The clutch comprises three largish eggs, more than double the size of the mallard's, each weighing about 13 per cent of the female's body weight. Incubation, as far as is known, lasts about 28 days. The young are evidently carried around, if need be, on the mother's back.

## THE SIZE OF BIRDS' EGGS

It is obvious that big birds like the ostrich lay larger eggs than small species such as the hummingbirds. Yet, considering their size, ostriches lay tiny eggs (each one is just 1.7 per cent of the female's body weight), while hummingbirds lay comparatively large eggs (17 per cent of their weight). This trend is common to most birds, and can be used to examine egg size as a survival factor.

In the case of ducks, big eggs mean fairly small clutches – an average of five eggs. Eight is about the norm for ducks that produce medium-sized eggs, and nine for those with smaller eggs. This pattern suggests that only so much energy can be put into reproduction; thus the choice must be between laying a few large eggs or a lot of small ones.

The principal advantage of a bigger egg is that it produces better developed offspring and also provides them with more food reserves when

*This female mute swan is incubating a large clutch of 11 eggs. These eggs have a rich yolk supply to nourish the young cygnets during their first few days after hatching.*

they hatch. The large egg of the blue duck contains a large amount of yolk, which is vital for the development of the embryo and for the survival of the chick for the first few days after hatching. Ducklings that hatch from big eggs are less dependent on the parents and can better withstand periods of food shortage.

The American black-headed duck is a brood parasite, laying its comparatively large eggs in the nest of coots. The newly hatched ducklings immediately leap out of the nest and fend for themselves. The young of diving ducks – goosanders, mergansers and scoters – must dive for food soon after hatching; in their case a big egg guarantees that they are well grown and ready to tackle the water when they are only a few hours old.

# BREEDING STRATEGIES OF AUKS

The auks are marine diving birds of the northern hemisphere which breed in Arctic waters around Alaska, Labrador and Newfoundland, Iceland, northern Europe and the USSR. Eighteen species are found in the North Pacific, six in the Atlantic, and three in both oceans. They range in size from the tiny least auklet, little bigger than a starling, to the duck-sized guillemots. All dive from the sea surface, to depths varying from 60 m (200 ft) to 180 m (600 ft), in search of small fish, sand eels, shrimps and planktonic crustaceans.

Most auks are social and breed in colonies, often in enormous numbers. The common guillemot breeds on both sides of the Atlantic and in the Pacific, at densities greater than those of any other bird. Yet each pair of birds knows the exact position of its tiny territory, and to avoid incubating the wrong egg, each bird's egg has a unique pattern and colour. Other colonial breeders include the puffins, the razorbill, the Pacific auklets and some of the murrelets; and of the few solitary species, the marbled murrelet breeds in trees in the temperate rain forests of British Columbia.

Newly hatched chicks display a wide range of development. Some, like those of the starling,

*The puffin. Its chicks emerge after a six-week incubation period, covered in a thick black down, but without the distinctive bill.*

hatch blind, naked and helpless. At the other extreme, the chicks of certain ducks and game birds are highly precocial, hatching with a good cover of down, able to maintain their own body temperature and to feed themselves immediately. The chicks of gulls are described as semi-precocial, hatching in a fairly advanced state but needing to be fed by their parents for some weeks.

Auks are unique among birds in the range of development patterns their chicks exhibit. The Atlantic puffin is a good example of what is referred to as a semi-precocial auk, laying a single egg from which a chick emerges about six weeks later, covered in dense black down. It is fed by both parents for another six weeks or so, and by then is almost fully grown.

*Breeding guillemots on Funk Island off Newfoundland. This species breeds at extraordinarily high densities with up to 70 pairs per square metre, yet each pair knows its own piece of rock.*

The ancient murrelet, a comparatively small Pacific auk, exhibits the precocial strategy. It is hunted by gulls and visits the breeding colony only by night. Unlike most seabirds, it lays two eggs, which are enormous in relation to its body weight (approximately 25 per cent of the female's weight). The newly hatched chicks are particularly well developed, and at just two days old leave the colony (again at night) and make their way to the ocean, guided by the light reflected from the sea's surface and their parents' individually recognisable calls. They are already capable of swimming and diving, but are cared for by the adults for several weeks.

Breeding behaviour midway between that of the puffin and the murrelet is displayed by the common guillemot, Brünnich's guillemot and the razorbill. They all lay a single egg, hatch their chicks after 32–35 days and feed them until they are about three weeks old. The quarter-grown chicks then leave the colony with their father who tends them for a further six weeks. The mother continues to visit the colony for a further week or so and then departs to spend the autumn and winter at sea.

*The common guillemot can dive to depths of 180 metres, the puffin only to 60 metres.*

Similar in many respects to the blue duck and Salvadori's duck is the handsome torrent duck of South America, represented by a number of races distributed along the Andes mountain range, each with different plumage patterns. But all have comparable lifestyles, inhabiting fast-flowing mountain streams and prising aquatic invertebrates from rocks with their narrow, flexible bills. Both sexes also possess a prominent bony spur on each wing, like the blue ducks, which provide extra support when clambering over wet rocks. The spurs increase in size as the ducks grow older, and the reason why those of the male are longer than those of the female – about 11 mm (½ in) and 6 mm (¼ in) respectively – is that his are additionally used for courtship display and for fighting rival males.

The plumage of the male torrent duck is very beautiful, with bold black and white markings on the head, brownish-grey body and wings, and a dark reddish-brown chest and underside. The female is much drabber, though both have red legs, a red bill and an iridescent green speculum or wing-bar. This marked difference in colour is unusual among river ducks, although common among ducks in general. Whereas in the majority of species the males moult into bright colours only for the breeding season, later adopting the plainer eclipse plumage, the male torrent duck retains his gaudy appearance throughout the year.

One pair of torrent ducks was watched daily by Jan Eldridge in Chile for over two months during the early part of the breeding season. Their territory extended more than 1,700 m (1,850 yd), but the birds did most of their feeding in areas of shallow water, the female spending more than twice as much time on this than the male, who stationed himself vigilantly close by. When not feeding, the pair searched the river bank for a suitable breeding site, accompanying this inspection with a set pattern of display postures and calls, and eventually choosing the old nest of another torrent duck. The three eggs were laid a week apart (not daily as with most other ducks). Each torrent duck's egg weighs 16 per cent of the female's body weight – a higher figure than for any other duck – clearly a strategy to produce a few very highly developed offspring.

*The distinctive male torrent duck of South America. Its long flexible bill is for probing amongst boulders underwater.*

The African black duck – our fourth river specialist – lives in the highland regions of southern Africa. As a rule it inhabits stony-bedded streams in woodland valleys, but occasionally it is found in open streams at altitudes of about 4,000 m (13,000 ft). It resembles the other river ducks in forming a long-term pair bond and in maintaining and defending a half to one kilometre stretch of water all the year round. The female normally lays a clutch of six eggs, incubating them alone. The ducklings hatch after about 28 days in a relatively advanced state of development. In contrast to the other river ducks, the female cares for them on her own by day, the male joining the brood to roost at night.

————————— □ —————————

The harlequin is also a river duck, but with a difference. It occurs in the Arctic, in Iceland, Greenland, on the east and west coasts of Canada, in north-eastern USSR and in Japan. Because it breeds so far north in such a markedly seasonal environment, its behaviour differs from that of the other river ducks of warmer latitudes. It cannot maintain a permanent territory, for in winter its breeding grounds are covered in snow and ice; and it does not form lasting pair bonds. It is compelled to spend the cold season at sea, eventually making its way to the mouths of rivers, streams and fiords and travelling upstream to breed. This behaviour resembles that of the salmon, and indeed bird and fish often share the same river.

It is a comparatively small duck with a shortish beak. The male is particularly handsome, mainly dark blue and deep red with white markings. The female has similar patterning but her coloration is less brilliant and at a distance she appears dark brown.

Harlequins dive for their food, propelling themselves underwater with partially opened wings. While breeding they feed almost exclusively on insects and their larvae, including those of caddis flies and mayflies, but especially of blackflies which lie beneath stones and in the gravel of the river bed. They also snatch adult flies from the water's surface. During the winter at sea, the ducks often hunt in flocks, eating snails, mussels and a few crustaceans, fish and worms.

Such a range of food, compared with that of other river ducks, is reflected in the small size of the harlequin duck's territory. Up to 50 pairs have been recorded in one 5 km (3 mile) stretch of river – with pairs only 100 m or so apart. Yet in regions where food is extremely plentiful, territories may not be staked out at all. Even so, the male usually defends a small area around his partner before the eggs are laid, presumably to prevent other males from mating with her. And because the territory is, in any event, only temporary, the ducks form a pair bond only for a single breeding season. The male plays no part in incubation or nest care, and the partnership breaks up as soon as the female lays her eggs.

In Iceland the harlequins nest on the ground, preferably in dense vegetation, only 5 m (16 ft) or so from the river. The average clutch is six eggs, each representing about 10 per cent of the female's weight. They are laid daily and the young hatch together after an incubation of some 28 days. Although they can feed themselves, they are tended by the mother, and at the age of about ten weeks are ready to fly off with her to sea, where they remain for the winter.

---

*The handsome markings of the male harlequin duck. This is another river specialist that spends its winters at sea.*

The four species of river duck which live in the southern hemisphere share a number of common features that are not exhibited by most other ducks. These features have developed because natural selection has operated in a similar manner on each species, moulding them to their special, white-water environment. It is a typical and fascinating example of convergent or parallel evolution, whereby animals – whether mammals, birds or reptiles – which may live in countries far removed from one another – show similar adaptations in order to cope with the demands of similar habitats. The key survival factors in the case of these geographically separated ducks are the establishment and defence of a permanent territory, the formation of a long-term pair bond, parental care by both sexes, and the production of a few well-developed offspring. And because of the specialised nature of their environment, often inaccessible and open to so many risks and threats, they deserve all the attention and protection that we can afford.

# UNDERWATER MAMMALS

During the brief summer season the polar seas are among the richest and most productive environments on earth. All through the long, dark winter months, in the waters locked under the ice, nutrients have accumulated; but light levels are so low that the planktonic plant life in the murky depths cannot utilise such resources. Come the summer, however, with its constant daylight, the plankton populations explode and provide the basis for a thriving food chain including fish, crustaceans and squid. Those creatures that can move into the polar regions during their short summers and crop their plentiful food resources are at an advantage.

These animals include the marine mammals – whales and dolphins, seals and walruses – in addition to a multitude of birds such as auks and penguins. There are 76 species of whales and 33 species of seals and their relatives, many of which spend part of their lives in polar seas. By and large, whales and seals are cold-water animals, but they do spend some time in warmer regions, and the related manatees and dugongs reside all the year round in tropical waters.

Although their tails move in different ways, the marine mammals, particularly the whales, superficially resemble fish. This similarity is no coincidence. It is another interesting example of convergent evolution, where the environment has shaped the creatures that live in it. Marine mammals have undergone extensive changes in body form to enable them to move easily in this medium. These modifications constitute a major survival factor. The most extreme alterations in body design have occurred among the whales since they spend their entire lives in water. Seals are less highly modified because they spend some of their time on land.

At 27 m (88 ft) or so in length, and weighing up to 150 tonnes, the blue whale is the largest animal that has ever lived. A land animal this size would need such enormous limbs to support it that it would almost certainly be incapable of moving. Buoyed by water, however, and equipped with fins, the blue whale can move with ease and astonishing grace. The smallest whales are the dolphins and porpoises: Heaviside's dolphin and the Gulf of California porpoise are both just 1.2 m (4 ft) long.

Most whales possess a streamlined, torpedo-shaped body with a long head, no neck and no hind limbs. Apart from the forelimbs, which have become flippers, there are few protruding parts to disrupt the flow of water along the body. Whales are almost hairless and their ears are reduced to tiny openings on the side of the head. Even the male's testes are internal, and the penis, which in some species is relatively long, is concealed within the body for most of the time. The tail flukes, consisting of two cartilaginous lobes, contain no bone, nor does the dorsal fin which certain species, such as the humpback, possess. The eyes are generally very small and in some whales vision is poor. Instead, they make extensive use of sound to navigate, and also to communicate with one another.

*A gray whale calf breaching.*

Opposite *A Weddell seal beneath a breathing hole in the frozen waters of the Antarctic.*

147

Whales propel themselves through the water by means of up and down movements of the flukes. They differ in this way from fish which use sideways flicks of the tail to drive themselves forward. The power of the whale's tail comes from its back muscles. The bigger whales usually swim rather slowly, but some of the dolphins, for short periods at least, may attain speeds of up to nearly 40 km (25 miles) per hour. This speed is achieved partly through the generated thrust of the tail but is also dependent on a further survival factor – their special skin. A boat, with its rigid hull, creates turbulence as it moves through the water, which slows it down. You can see this in the bow waves and the wake of a ship. The dolphin's flexible covering of skin and blubber – a layer of fat just beneath the skin – helps to minimise this turbulence. Moreover, the outermost layer of skin produces waxy secretions which further reduce disturbance in the water and increases swimming efficiency. This lubricating action is similar in principle to skiers waxing their skis to make them move faster through the snow.

The diving ability of whales is remarkable. The deepest dives recorded have been made by sperm whales, presumably in search of squid, their main prey, which lives at immense depths. These whales may plunge as deep as 2,000 m (6,500 ft) and remain submerged for more than an hour.

*The tourists enjoy a rare opportunity to see a gray whale at close quarters.*

## BREATHING AND BUOYANCY IN WHALES

Whales, like all mammals, breathe air, and between dives they must surface to expel the de-oxygenated air from their lungs and replace it with oxygenated air. The cry of 'There she blows!' yelled by the old whalers referred to the cloud of spray produced by a whale when it was exhaling. Exactly what the 'blow' or 'spout' consists of has long been a mystery: it now appears that it is a mixture of water vapour and some special secretions from the windpipes. The spout is often the first sign of a whale's presence, and many species can be identified by experts from its size and shape.

Whales, like seals, avoid the 'bends' – caused by nitrogen diffused in the blood – by carrying relatively little air in their lungs when they dive, so reducing the nitrogen level. And the sperm whale has an additional, unique survival factor, linked with diving and buoyancy. The whale has a disproportionately large head, most of which is taken up by a wax-filled 'spermaceti organ'. For the whaling industry, spermaceti was important for producing a high-quality oil used as a lubricant and in the manufacture of cosmetics. The spermaceti organ is interesting for the part it is believed to play in controlling the speed of the whale's dive. The whale can apparently regulate the temperature of the wax, which melts at 29°C (84°F), by moving it in and out of the nasal passages and sinuses, where the blood flow is rich. To cool the wax, the whale inhales a quantity of cold sea water. During a dive, the wax may cool and so shrink and solidify. This increases the density of the head. Some people think this may permit the whale to make a very

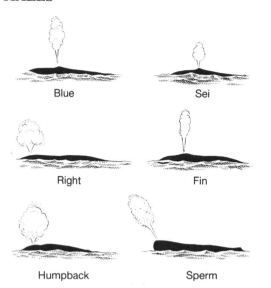

*Whales can be distinguished by the shape of their 'blow' at the surface.*

rapid descent – up to 170 m (285 ft) per minute – although this is not yet proved. During the ascent, the wax warms, increasing the whale's buoyancy and helping it to rise quickly to the surface.

Other toothed whales possess a smaller wax-filled body inside their head, known as the 'melon', which probably serves to focus sound during echo-location. In dolphins it is thought that high-frequency sounds, generated and projected from the melon, are actually used to stun and kill prey, but this has still to be confirmed.

Marine mammals, living as they do mainly in water, face two main problems. They must breathe air and they must keep their body temperature constant. Most whales maintain their body temperature at 36°–37°C (97°–99°F), which is the same as a human's, even though they are in water which is invariably cooler and sometimes close to freezing. The answer to how they do it lies in a number of specialised survival factors.

Firstly, their large size and streamlined body helps to minimise heat loss: a large body, with its small surface area relative to its volume, loses heat less rapidly than a smaller one. Secondly, whales are remarkably well insulated. Many Arctic mammals, such as the polar bear or musk ox, have a

149

thick coat of fur to insulate them. But fur would make whales inefficient swimmers, and being wet would not keep them very warm. Instead, whales insulate themselves with blubber. In some species this layer of fat can be quite thick – about 50 cm (20 in), for example, in a 16 m (53 ft) long bowhead. The amount of blubber present varies throughout the year. During the summer many whales visit the productive polar seas in order to feed and build up their fat reserves; they then use these reserves in winter which is spent in warmer waters. Whale blubber is a smelly, rubber-like substance, and the beaches around Inuit (Eskimo) settlements in the far north are often redolent with its pungent and characteristic odour.

A thick layer of blubber is extremely effective in keeping out the cold, and also for keeping the body heat in. Whalers sometimes used to exploit this fact by cutting deep inside a whale to get at the meat which was virtually cooked by the internal build-up of heat. This does not happen when the whale is alive because its circulatory system keeps the body at blood temperature. As soon as the whale dies, however, the circulation breaks down and the heat is retained long enough to 'cook' the meat.

*The killer whale has an impressive set of teeth.*

Maintaining the body temperature at a relatively high and constant level (compared to the sea) allows the various physiological processes to operate efficiently at all times, and is one of the important keys to a marine mammal's success. To achieve this it needs to consume large quantities of food. Whales take in an extensive range of food types and display a variety of adaptations for feeding.

Whales can be divided into two main groups, those with teeth and those with baleen or whalebone. The toothed whales include dolphins, porpoises, sperm whales and beaked whales, while the baleen whales

include the gray whale, rorquals and right whales. The diet and feeding methods of these two groups differ dramatically. The toothed whales, ranging in size from the dolphins to the sperm whale, evolved from a terrestrial carnivore, but in the course of evolution the jaws of many of them have become elongated to form a beak. The ancestors of these whales possessed teeth of different shapes – incisors, canines and molars – but these are not suitable for catching and eating fish and squid, the principal prey of whales. The majority of toothed whales have a set of similarly shaped, sharp, pointed teeth. The actual number is sometimes less than in other mammals; indeed some of the beaked whales, such as the strap-toothed whale, have only a single pair of teeth.

Dolphins are mainly hunters of fish and squid, but the largest of them, known as the killer whale, is a highly efficient predator of seals, whales or penguins. Killer whales usually hunt in packs and live in permanent groups of about 20 animals. Each group consists of one adult male, readily distinguished by his large size and huge dorsal fin, three or four adult females and a number of immature animals. These voracious predators have been known to knock penguins off ice floes in order to reach them, and almost to beach themselves in their pursuit of seals. They have also been seen to attack and kill a blue whale – the largest baleen whale.

*Killer whales normally hunt and live in groups known as pods.*

151

*The 4 m long jaws of a gray whale which has washed up on the shore. The baleen plates, which act as filters when the creature is eating, are visible here.*

The baleen whales have a large head with enormous jaws and several hundred baleen plates, hanging in two rows from the roof of the mouth, which act as filters. Baleen is a horny substance, also known as whalebone. The jaws may be 4 m (13 ft) long in a bowhead, and in all species they are capable of opening extremely wide. The 'pleats' on the chin and throat of baleen whales swell as they take into their mouth a huge quantity of water which contains small crustaceans, especially krill, which are like shrimps, and small shoaling fish such as capelin. The water is then expelled through the baleen plates, which sieve out the food animals.

Most of the baleen whales are monogamous, each male mating with a single female. As a rule the most dominant male will assert sole rights over the partner of his choice. Among the southern right, bowhead and grey whales, however, competition between rival males can be fiercely aggressive, for the females of these species routinely mate with more than one male. In multiple matings the sperm from the different whales compete within the female's reproductive tract to fertilise her egg. This competition may explain why the males of these species have disproportionately long penises and large testes: the closer the male can get his sperm to the female's egg, the more chance the sperm will have of fertilising it, because it is given a 'head' start. The toothed whales, on the other hand, are generally polygamous, with one or more mature males defending a harem of females. As with the baleen whales, mating probably takes place near the water surface, each animal in belly contact, either in a vertical or horizontal position.

The majority of whales have clearly defined breeding seasons. The females give birth to a single offspring, following a lengthy gestation period. Whales do not breed until they are aged five–ten years or more. The larger species may then reproduce only once every two years whereas the smaller dolphins may breed annually. Killer whales reproduce only once every three to eight years. But whales are long-lived, and can afford such slow breeding rates. The smaller ones live up to 40 years, while the blue, sperm and killer whales may live for as long as 60 years.

———————————— □ ————————————

Over the centuries whales have been hunted, mainly for their blubber and oil but also for meat. The plankton-feeding species were additionally valued for their baleen, formerly used in the manufacture of corsets and umbrellas. Human exploitation of whales has proved disastrous for some populations; the northern right whale and the blue whale have been hounded and now occur in very low numbers.

In 1982 the International Whaling Commission banned commercial whaling. Yet despite this certain nations have continued to kill whales – legally – by exploiting various loopholes in the Commission's small print. Japan, for example, persists in catching whales under the pretence of 'scientific whaling'.

For those of us who refuse to regard them as a potential source of commercial profit, whales inspire a sense of admiration and awe, particularly in the wild. Once, while studying marine birds on a group of

*Killer whales produce only one offspring at a time.*

*Although whaling has been significantly reduced in recent years, some nations such as Japan continue to slaughter whales.*

remote islands in Labrador, drifting pack ice moved in and completely surrounded us. The ice stretched from horizon to horizon with barely a break; there were just a few patches of open water between the islands. In one of these ice-encircled areas of water three minke whales suddenly surfaced. Eventually one of them stood up on its tail flukes with its entire head pointing vertically out of the water, and looked around. It did this several times before sliding back under the surface. All three then disappeared. Whether they got out of the pack ice to open water, we will never know.

On another memorable occasion we swam with a bottle-nosed dolphin off the coast of Pembrokeshire. This dolphin, subsequently to become famous on television as 'Donald', was completely wild but clearly enjoyed the company of humans and small boats. On this day he appeared on the north side of Skomer Island, but vanished as we hurriedly put on our wet suits and swam out into the bay. He soon reappeared, however, swimming very fast towards us with just his dorsal fin showing, looking uncomfortably like something out of *Jaws*. He stopped a few feet in front of us and opened his mouth – revealing a fine set of sharp teeth – then turned and made it clear that he simply wanted to play. He allowed us to ride around on his back and swam over and under us in a boisterous game of rough and tumble.

Dolphins will sometimes give assistance to injured swimmers by pushing them to the surface, so enabling them to breathe. This does not necessarily mean that they have a special feeling for humans since they normally tend members of their own species, particularly the young, when they are in difficulties. But it does help to strengthen our relationship with these attractive, intelligent animals. Moreover, it is possible that we will come to look upon them as more than dolphinarium entertainers. As an experiment hospital patients suffering from depression were encouraged to swim in company with dolphins in an attempt to alleviate their condition. Many of the patients, when interviewed, said that the experience had helped them enormously.

*Opposite A bottle-nosed dolphin.*

Like the whales, seals have a streamlined body shape, with no obvious neck and few projections. Seals likewise have an insulating layer of blubber but differ in having a covering of hair, which in some species, like the fur seals of the polar regions, can be quite luxuriant.

The seals fall into three broad groups. The eared seals, such as the fur seals and sea lions, have external ears and can tuck their hind flippers forward to 'walk' on land. The true seals, which include the harp, grey and elephant seals, have inconspicuous ears and hind flippers which cannot support the weight of the body, so they have to hump themselves along on land. Finally there is the largest seal of all, the walrus, which is closer to the sea lions, but which is so heavy that it heaves itself around more like a true seal.

All three groups of seals swim in rather different ways, as is dictated by their anatomy.

The power source in eared seals is located in their large foreflippers which enable the animals to 'fly' through the water, the hind pair playing no part in swimming except perhaps as a rudder. On land, however, the hind flippers of the eared seals permit them to move much more efficiently than other seals. These species also possess very large neck vertebrae for muscle attachment.

The true seals use their hind flippers together with undulating body movements in order to propel themselves through the water. Consequently it is the lumbar vertebrae which are enlarged. The foreflippers of true seals are employed mainly as paddles during relaxed swimming. The hind limbs are all but useless on land.

The walrus is much more ponderous than either of the other two groups of seals. It swims by using both its foreflippers, like eared seals, and its hind flippers, like true seals.

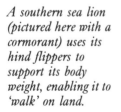

*A southern sea lion (pictured here with a cormorant) uses its hind flippers to support its body weight, enabling it to 'walk' on land.*

Herculean feats of diving are not unexpected of enormous sea mammals like the sperm whales. Yet some of the much smaller diving animals, such as seals and penguins, are also astonishingly deep divers. The Weddell seal, which breeds farther south on the ice of Antarctica than any other seal, feeds on fish, squid and crustaceans which it can obtain only by deep diving. Individuals have regularly been seen bringing deep-water fish that weigh 30 kg (66 lb) up to the surface. Because their ocean habitat is frozen for much of the year, the seals have evolved a sophisticated survival technique. Pregnant seals return to Antarctica in late August under the sea ice, to the stable areas of ice near the shore, where it is not more than 2 m (6½ ft) thick. When a female finds a crack under the ice, she grinds away at it with her specially adapted teeth. The front teeth, canines and incisors, are strong and point outwards. Soon she will have made a hole big enough to climb up on to the ice. In a few days she will give birth to her pup. As the pup grows, the mother goes fishing by dropping down through the hole and diving to the depths. Each time she returns she inspects the hole. If it

*Three ponderous walruses. Those enormous tusks are used in fighting for mates, and sometimes to heave themselves out of the water.*

157

# DIVING BIRDS

A number of birds take advantage of the wealth of food below the water's surface. These include a variety of diving ducks, such as the eiders, the sawbill ducks (like the goosander and the smew), grebes, divers, auks and penguins. All these species possess several survival factors that allow them to exploit underwater niches. The auks share certain common features with the penguins, their southern hemisphere counterparts, but differ in one vital respect; whereas all penguins are flightless, all living auks can fly. The extinct great auk, however, once known as the penguin, was equally big and similarly flightless.

The wings of penguins have gradually been reduced in size to act as flippers or paddles. The auks also have fairly small wings, but because they have retained their capacity for flight, the wings have not been reduced quite so much. The auk wing represents a compromise between aerial and underwater flight. In the air these birds, and particularly the largest species such as the common guillemot, must beat their wings very rapidly in order to remain airborne. In the sea the wings are partially closed but still beat, and the auks seem to 'fly' underwater. In contrast both to auks and penguins, the grebes and certain ducks use their feet for propulsion underwater. The goosander is very versatile; it uses both feet and wings.

The penguins have specially modified, much reduced feathers, and rely on a thick layer of subcutaneous fat for insulation, much like the blubber of whales and seals. The auks, however, although possessing a fatty layer beneath the skin, keep warm mainly as a result of their dense plumage. Guillemots can maintain their body temperature even when the surrounding air temperature is as low as −55°C.

The presence of a bulky covering of feathers probably limits the depths to which auks can dive. The greatest depths are achieved by the biggest species; the common guillemot dives to a maximum of 180 m (600 ft) and the smaller Atlantic puffin to a mere 60 m (200 ft). The champion diving bird is the emperor penguin. Individuals fitted with dive meters have been recorded as descending to 300 m (1,000 ft), remaining submerged for over 15 minutes and swimming underwater for a distance of 360 m (1,200 ft) between two ice holes!

*The goosander dives beneath the water's surface to catch fish with its 'saw' bill.*

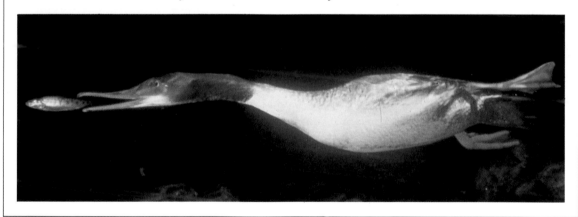

shows signs of freezing over, she will gnaw away at the edges, in some cases with such force that her gums bleed. Without the hole, she cannot suckle her youngster and it will die.

The mammary glands of the female seal are spread over her underside and neither they nor the nipples protrude, except briefly when suckling. Similarly, the male's reproductive organs are tucked away so that they do not increase drag in the course of swimming. In most male mammals the testes are housed in a scrotum, outside the body cavity. This keeps them cool in relation to the temperature of the rest of the body, and allows the production of viable sperm. The testes of seals, despite being internal, are maintained at an appropriate temperature through an ingenious arrangement which involves their being supplied with cool blood directly from the hind flippers.

Because they have no predators on land, Weddell seals are fairly tame and approachable. The American scientist Gerald Kooyman studied the seals under natural conditions in the Antarctic. Individual seals fitted with depth gauges were found to remain underwater for up to 15 minutes at a time, with dives down to about 100 m (330 ft). However, the longest dive recorded lasted 43 minutes and the seal reached a depth of 600 m (2,000 ft). Like other marine mammals, Weddell seals avoid decompression

*A Weddell seal.*

sickness, or 'the bends', by diving very rapidly, especially beyond depths of 50 m (165 ft) or so, with only partially filled lungs. On returning to the surface, the seal breathes heavily to ventilate the lungs completely. For a deep dive it may need up to an hour to recover, getting rid of the waste products that have collected in the body without oxygen from the air. Apart from the physiological ability to dive to great depths, a further survival factor is their skill in navigating under the ice. Kooyman's studies showed that the seals could swim directly and unerringly between holes more than 1.6 km (1 mile) apart.

Seals retain body heat by means of both blubber and fur. In land mammals a coat of fur provides effective insulation because it traps a layer of stationary air against the skin. In underwater mammals, the insulating layer of air is dissipated or diminished as the fur becomes wet, but some reduction of heat loss is achieved, particularly in the fur seals where the fur

# MANATEES AND MERMAIDS

The Sirenia, comprising the three species of manatee and the dugong, differ from whales and seals in one important respect: they are vegetarians. They are all about 4 m (13 ft) long when fully grown, slow-moving and, like other marine mammals, insulated with a subcutaneous layer of blubber. Their leisurely and graceful movements, though hardly their looks, almost certainly gave rise to the myth of the mermaid.

Whereas the manatees are either river dwellers or confine themselves to coastal waters, the dugong is truly a marine mammal. Its tail is much more like that of a whale, with distinct flukes, while the manatee's consists of a single rounded paddle.

The diet of these animals comprises submerged or floating plants, and their teeth are modified accordingly. The dugong's canine teeth on the upper jaw project forwards like a pair of small tusks. They are used to dig for the nutritious rhizomes of aquatic grasses. In the manatees the molar teeth are replaced in an unusual horizontal manner, with those at the back of the mouth 'migrating' steadily forwards at a rate of about 1 mm per month. As their roots become exposed, they simply fall out at the front. The Caribbean and African manatees eat turtle grasses which grow in shallow water. The Amazon manatee feeds on floating vegetation. But when there is little plant growth the manatee lives off its blubber.

Like whales, manatees and dugongs are long-lived, with a slow breeding rate. Gestation lasts 12 months, and the mother gives birth to a single offspring. The female manatee suckles her young, sometimes hugging it to her breast with one flipper, and tends it for up to two years. It is able to breed at about six years of age.

Manatees are frequently killed for their flesh; they are slow swimmers and are easily speared. All the manatees are presently endangered. Dugongs are currently threatened by the increasing use of strong nylon fishing nets from which the creatures cannot break free; they are often caught and drowned in these. The dugong's closest relative, Steller's sea cow, which was also tasty to eat, disappeared two centuries ago. It was discovered in 1741 and hunted to extinction within 30 years. The animal lived only in the cold waters of the North Pacific around Bering and Commander (Komandorskiye) Islands, feeding on kelp – brown seaweed. The survival factor that enabled them to cope with their low temperature environment was their huge size: sea cows were three times the size of the dugong and manatee.

*The slow-swimming manatee or sea mermaid.*

is unusually dense and the skin is richly endowed with oil-producing sebaceous glands. The oil keeps the underfur waterproof, so that when the seal dives, some air is trapped and water is prevented from reaching the seal's skin.

*Squid are prized prey for many underwater mammals.*

This system still has certain drawbacks. One is that the deeper the seal dives the more compressed the air layer becomes and the less effective is the insulation it offers. The other problem is overheating. An actively swimming seal generates a lot of body heat and this is retained by fur. So heat can be lost only through the hairless flippers. Because of these limitations, blubber is the seals' main type of insulation.

———————————— □ ————————————

Seals, like whales, feed on a wide range of prey. Some will eat almost anything they can catch, others are specialists.

Despite its name, the crabeater seal, like the ringed seal, feeds on krill. These two species do not have baleen plates for filtering these minute crustaceans: they possess highly modified teeth, with several cusps, through which they sieve their prey. The southern elephant seal and Ross's seal consume squid, while the walrus and the bearded seals feed principally on bottom-dwelling molluscs.

161

The 3 m (10 ft) long Antarctic leopard seal is an aggressive predator which feeds on other seals and penguins in addition to fish and krill. Leopard seals lurk in the waters beside penguin rookeries, waiting for a meal to arrive. In an effort to avoid this long, sinuous hunter, penguins return to their colonies in groups, swimming at high speed.

Leopard seals are not only fast swimmers but can also move unexpectedly quickly on land as well. A colleague discovered this on his first trip to the Antarctic when he came across a sleeping leopard seal on a beach. Recalling that this species is cumbersome on land, he approached confidently to take its picture. The seal awoke and, contrary to expectations, chased him rapidly across the beach. He escaped by a hair's breadth. Judging by the 30 cm (12 in) scars found on some crabeater seals, the leopard seal's main prey, his escape was a fortunate one.

———————————— ☐ ————————————

Seals have to return to land, or ice, in order to breed. Many northern true seals and the walrus breed on ice, although this is not true of the eared seals. Because seals are often clumsy on firm or slippery surfaces, and vulnerable to predators, they generally choose rather inaccessible locations. This, in turn, has a profound effect on their mating patterns. In most species, including the two and a half tonne elephant seal, beaches suitable for giving birth are few and far between. Females therefore tend to congregate on those available and may be monopolised by just a few males. In California bull elephant seals arrive on the beaches before the females and compete among themselves for dominance, inflating their trunk-like protrusions as they threaten and fight each other.

The largest and strongest bulls gain top position in the hierarchy. The competition to become alpha male is intense, and the benefits are considerable. The most dominant bull may mate with over a hundred females in a season, but to achieve this he will need to engage in many savage battles with his rivals, during which time he will have to go without food. If he were to leave the beach to feed at sea, other males would inseminate 'his' females.

The female elephant seals appear on the beaches in January and drop their single pup soon afterwards. Only a month or so later, when the pup is being weaned, do the females mate – usually with the harem master. But this is not always so because subordinate bulls employ a variety of devious tactics in the struggle to father some young. Small males, which have small trunks, come ashore quietly, exploiting their female-like appearance to slip past the dominant male. They then sidle up to a receptive female and attempt to mate with her. It is probably not in her interest, however, to copulate with a subordinate male, so she usually lets out a cry, attracting the attention of the dominant male, who rushes over. Should the subordinate male not retreat, he runs the risk of serious injury. But despite

the hazards, one or two of these males that sneak in to the harem will mate successfully in this way.

The breeding behaviour of the ringed seal is different. They breed on ice, with the females constructing a den under a crust of snow. These dens are conspicuous and probably serve to keep the pups warm rather than safe from predators. In one study no less than 58 per cent of newborn pups were spotted and killed by predators, mainly polar bears. In the Canadian Arctic, pups are born during April and suckled for about six weeks. Then the females mate in the water under the ice.

Although seals, like whales, have been, and continue to be hunted, their future appears much brighter. Despite massive exploitation in the past, careful protection has resulted in large increases in the populations of once endangered species like the elephant seal, the walrus and the New Zealand fur seal, which are back from the brink.

*A pair of mating elephant seals. The bulls will fight savagely to hold and mate with a harem of females.*

163

# WOODPECKERS

In healthy woodland most of the trees grow gradually but steadily, over the years becoming taller and broader, their branches thrusting out to touch those of their neighbours until they form a virtually continuous canopy overhead. Yet trees, despite their illusion of permanence, do not last forever. As time passes, parts of the tree begin to die. Branches may be damaged in storms, the bark may be attacked by fungus and whole trees may die prematurely from disease. Whatever the cause, the dead portions of the tree slowly decompose. This decaying process, which may take years, is controlled by a number of organisms; and the principal roles are played by fungi and insect larvae.

Dead branches often provide suitable nest sites for birds; they also contain an abundant supply of nutritious grubs, provided the birds can reach them. While the many insects which feed on the leaves of growing trees move about visibly and are easily accessible to birds, these grubs are buried deep inside the rotting wood of the dead or dying tree. Since they cannot be seen with the naked eye and make no noises to betray their whereabouts, they might seem to be safe from predators. However, woodpeckers have developed very specialised methods of getting at these grubs, some of which, like the longhorn beetle's, may stay in the wood for several years and grow very large. Woodpeckers are among the few birds that can make a living in deciduous woodland all year round, although survival is more difficult in the winter.

Woodpeckers have evolved a number of important survival factors. These include a strong, pointed bill designed for hacking and hammering bark or wood, a reinforced skull to protect the brain from being damaged

Opposite *The lesser spotted woodpecker of Europe is little larger than a sparrow.*

while doing so, and a long, protrusible tongue for teasing insects out of holes. These activities necessarily involve climbing, and most woodpeckers are able to hop around nimbly by using their feet in conjunction with their tail. The feet have two forward-pointing and two backward-pointing claws, and the tail feathers are exceptionally rigid, providing a prop when climbing vertical tree trunks.

Many birds nest in tree holes but usually only in those that are ready made. Woodpeckers, however, are capable of excavating their own nest chambers, either in live or dead wood. There are 198 species of woodpeckers distributed throughout the world. Of these, 169 are known as true woodpeckers; among the rest are the piculets (27 species) and the two species of wryneck. The diminutive piculets, up to 12.5 cm (5 in) long,

*A pileated wood-pecker hammers energetically at a tree trunk.*

are a poorly known group of birds which inhabit tropical parts of America, Africa and Asia. The wrynecks, which are rather like primitive wood-peckers, live in Europe and Africa. The true woodpeckers occur in all parts of the world except the polar regions, and are notably absent from Australia. In that country the role of dead-tree forager is assumed by the treecreepers; but these birds forage only on the surface of trees and do not dig deeply into dead wood, like the woodpeckers. The reason why there are no woodpeckers in Australia may be explained by the fact that the ubiquitous eucalyptus trees do not rot in the same manner as other trees, thus failing to provide sufficient decaying wood for the birds to exploit.

Woodpeckers are most abundant in south-east Asia and South America where as many as 13 species may inhabit the same area of forest. But

woodpeckers, wherever they live, generally avoid competition for living space and food. They do this by having their own lifestyles and ecological niches – meaning that they display different habits and feed on different trees or different parts of a tree. This explains why woodpeckers that live together may vary considerably in dimensions. North America possesses over 20 species. Sizes range from the pileated woodpecker (37.5 cm/ 15 in), a close relative of the almost extinct ivory-billed woodpecker, through medium-sized ones like the hairy woodpecker (17.5 cm/7 in), to the small downy woodpecker (12.5 cm/5 in).

Exactly the same principle occurs among the three British woodpeckers; the green woodpecker is the largest (36 cm/14 in), followed by the great spotted (23 cm/9 in) and the lesser spotted (14 cm/5½ in) woodpeckers.

*Nuthatches may nest in old woodpeckers' holes, sometimes plastering the entrance hole with mud.*

## THE WRYNECK

The two species of wryneck, one European, the other African, are closely related to the true woodpeckers but do not possess stiffened tail feathers, a strong bill or a reinforced skull – the adaptations associated with hammering wood. But they do have a greatly elongated tongue for capturing ants; and their plumage, unlike that of most woodpeckers, is fairly drab, providing them with superb camouflage as they forage on the ground or in trees.

Wrynecks feed almost exclusively on ants – adults, pupae and larvae – either picking them directly from the ground or using the protrusible tongue to winkle them out of crevices. When feeding their young, the adults sometimes bring a strange assortment of indigestible objects to the nest, including bits of plastic, stones, bones and snail shells. The last two items may provide the chicks with calcium, likely to be deficient in a diet consisting largely of ants; the stones may help to grind up the stomach contents, and the plastic is perhaps picked up in mistake for the other objects.

The wryneck's name is derived from its habit of twisting its neck in snake-like fashion while feeding, although the purpose of this contortion is not known. However, it is interesting that the bird hisses loudly like an angry snake when threatened. Furthermore the wryneck will also feign death if cornered – usually after a bout of hissing and head-twisting – closing its eyes and hanging limp.

In Britain the wryneck was until recently a summer visitor. Its last foothold was in Scotland, where its favourite habitat was open woodland. Its decline in Britain, and indeed over western Europe, is thought to be due to cooler summers reducing the numbers of its ant prey.

With its smart black and white plumage and brilliant crimson-red head and tail-covert feathers, the male great spotted woodpecker is unmistakable. Its 'pick' call, often uttered from the topmost branches of a tall tree, and its mechanical drumming, are equally distinctive. Indeed, it is one of the best known of all woodpeckers and it typifies the special habits and adaptations of so many of its relatives.

The great spotted woodpecker is one of the most abundant and widespread of European species, with a range that extends over most of the continent eastwards through Russia as far as China and Japan. Curiously, all woodpeckers are absent from Ireland, but so too are several other British animals, such as the grass snake, the adder and the common toad. There is no agreed explanation for this

Male and female great spotted woodpeckers are similar in size and plumage, but the male can readily be distinguished in the field by the red patch on the back of his head. Moreover, depending on where they live, the birds may differ in body dimensions and the proportionate size of the bill. Individuals in southern France are small with a relatively long beak, while those further north in Scandinavia are bigger with a fairly short beak.

These two patterns, a larger body size and a decrease in beak length at higher latitudes, occur in a number of birds, and similar principles apply to body size in other orders of animals. Although there are some exceptions, they are sufficiently widespread to have been summarised as 'rules'. Bergman's rule states that animals living in cold climates tend to be bigger and heavier than those occupying warm regions; and according to Allen's rule, animals adapted to cold have smaller body extremities (ears, bills, etc.)

*Opposite A juvenile great spotted woodpecker, with down on its flanks and distinctive red cap.*

than their counterparts in warmer countries. However, this is a very generalised rule, and diet may also be a partial cause of varying dimensions. The woodpecker's smaller beak in northern Europe may be connected with its bias towards seed-eating.

The numbers of great spotted woodpeckers are actually increasing at the present time. During the 19th century there was a marked decline in Europe, probably because of extensive tree felling and competition with starlings for nest holes. But since the beginning of this century they have made something of a comeback. In fact, since the 1960s their numbers have increased rapidly, and during the 1970s their populations doubled, apparently as a result of Dutch elm disease.

This disease has killed millions of elm trees and scarred many a landscape. It is caused by a fungus and is spread by beetles, or through the roots of neighbouring trees. Its total effect on British wildlife is difficult to evaluate. Some bird species, such as the robin, goldcrest and dunnock,

*The unmistakable signs of Dutch elm disease. It is caused by a fungus and spread by a beetle which burrows under the bark.*

have suffered from the disappearance of elms, which probably provided them with a nesting habitat and an important source of food, either directly as seeds or insects on the leaves, or indirectly in the form of ground-living invertebrates under the tree canopy. However, other birds, like the great spotted woodpecker, have benefited.

*The black wood-pecker is the largest of the European woodpeckers.*

Two possible reasons for the dramatic recovery of the great spotted woodpecker population in Britain may be an increase in the number of potential nesting sites and the quantity of available food. The wood of dead trees becomes soft and easily excavated only if it is subsequently attacked by fungi, especially the heart-rot types such as beefsteak fungus and sulphur polypor fungus. But since there is no evidence that woodpeckers have suddenly switched to nesting in elms, it is clear that elm disease has not created a lot of new nesting sites. The more likely explanation, therefore, is an augmented food supply.

Scolytid beetles, the vectors of Dutch elm disease, can occur at very high densities in infected trees, and their grubs form a tasty meal for a woodpecker. An adult burrows through the bark and digs an egg chamber. The larvae that hatch from these eggs then excavate galleries as they feed, radiating out from the chamber and producing the sculptured patterns seen on the inside of bark from diseased elms. If the adult which lays the eggs is carrying the spores of the heart-rot fungus on its body, the fungus spreads along the larval galleries. When the larvae change into adult beetles and dig their way through the bark, they carry the fungus with them to the next elm. The beetle larvae can be extremely numerous: one square metre of trunk may accommodate as many as 5,000, and a single elm may contain a million larvae. Since great spotted woodpeckers eat both the adult beetles and their larvae, this clearly represents a substantial food source.

The years of plenty for these woodpeckers may nevertheless be fairly short-lived. Although Dutch elm disease continues to affect trees, the number of elms left in Britain is now small. Once the tree has died, the bark remains suitable for the beetles for just two years; after this time it becomes rotten and falls off, leaving the bare skeleton of the tree. Additionally, in some areas all dead elms are cut down and burnt.

———————————— □ ————————————

The great spotted woodpecker feeds on insects, including caterpillars and grubs, in summer, and on grubs and tree seeds, including pine cones, in winter. Because of its specialised adaptations, it can exploit a winter food supply unavailable to other birds. Like other woodpeckers, it has a stiff tail for additional support when climbing. Its twelve tail feathers are arranged in pairs, and the three innermost pairs have a very strong shaft and sharp ends. The bird holds these feathers firmly against the trunk of the tree as it climbs, and is therefore balanced on three points – the two feet plus the tail.

When foraging, the great spotted woodpecker makes exploratory taps with the bill, presumably to test the consistency of the wood beneath the bark. The bird also looks for visual clues, preferring to feed on branches which display lots of insect holes. Many insects that live in wood dig tunnels or galleries, just like the *Scolytus* beetles in elms. To get at these the woodpecker uses its bill and its specially adapted tongue. The strong bill is used to rip the bark from the tree or to hack into the wood to a depth of about 10 cm (4 in). The long, narrow tongue, extending 4 cm (1½ in) beyond the end of the bill, has a harpoon-like tip with which it can impale prey. The tongue also has bristles which are covered with sticky saliva so that any insects touched are instantly trapped.

During spring and summer the main diet is caterpillars, but if these run short, then a more macabre side of the woodpecker's nature appears – as a predator of other birds' eggs and nestlings. If desperate, the great spotted woodpecker will raid those species that lay eggs in tree holes or in nesting boxes. Great and blue tits, redstarts, pied and spotted flycatchers may have their nests ransacked and their eggs or young eaten. Species like the willow tit, which generally build a nest in a rotten tree stump, are especially vulnerable; in some areas woodpeckers may destroy more than half the nests constructed by willow tits. The birds may also attack nest boxes set up for the study of species such as the blue tit. The woodpecker quickly homes in on the box, either enlarging the entrance or hammering a hole straight through the side. If the young tits are close to fledging, there may be no need even to peck the box open. Hearing the predator alight, the young assume that it is one of their parents returning with food, cluster around the nest entrance hoping to be fed and are immediately snatched up in the woodpecker's powerful beak. Both adult tits and pied flycatchers clearly recognise woodpeckers as enemies and mob them furiously if they venture anywhere near the nest.

Early in the spring the woodpeckers may drink the rising sap from trees

*A juvenile green woodpecker takes a look at the outside world.*

# BIRD TONGUES

The tongues of birds are seldom visible, yet they perform a range of functions which are essential to survival. The tongue is fleshy and its movements are controlled by a pair of hyoid bones at its base. The tongue has nothing to do with the utterance of bird song or calls, which are produced by a special organ, the syrinx, situated in the windpipe. Nor is it primarily associated with taste. The fact that the tongue, except among parrots, contains few taste buds, indicates that this sense is poorly developed in most birds.

The tongue is therefore used almost exclusively for obtaining food; and its proportionate size and structure reflect the feeding habits and diets of the birds concerned. Storks and king-fishers have very short tongues, relying mainly on their large beaks to grip aquatic prey.

*The tongues of birds vary considerably from species to species. Pictured below are: (1) wryneck; (2) white-headed woodpecker; (3) red-breasted merganser; (4) bananaquit; (5) shearwater.*

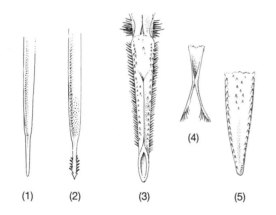

(1)   (2)   (3)   (4)   (5)

Penguins and auks, too, feed on fish, but they both have long tongues; that of the penguin is covered with small, pointed projections which help to grasp the slippery prey, whereas the auk's tongue is smooth, compensated by a very rough roof to the mouth.

The tongues of woodpeckers vary in length and texture. The very short tongues of the sapsuckers are tipped with fine hairs. This is a survival factor that enables the birds to collect sap from trees by capillary attraction, rather like a paint brush absorbing a drop of water. Some woodpeckers have protrusible tongues that can extend way beyond the tip of the bill – over 10 cm (4 in) in the case of the green woodpecker. It is not that the fleshy part of the tongue is markedly longer than that of other birds, but that the hyoid bones, which operate it, are greatly extended. So large are they that in some species they reach over the back of the skull and are attached to the upper part of the beak; and in others they are wrapped around the bone that encircles the right eye. These long tongues are often barbed and covered in mucus in order to extract insects from holes and crevices. The longest tongue of all is that of the wryneck, which specialises in eating ants – 16.5 cm (6½ in) – nearly two-thirds of the bird's body length. There is an interesting parallel here with the anteaters, mammals which have also developed extremely long tongues used for the same purpose – one of many instances of convergent evolution in the animal kingdom.

The tongues of hummingbirds and sunbirds, which feed mainly on nectar, are long and folded over to form a tube. The birds force nectar into the tongue by thrusting it rapidly in and out of the flower.

– a habit that is rather rare in Britain but common elsewhere in Europe. In this they emulate one group of North American woodpeckers familiarly known as 'sapsuckers'. The great spotted woodpecker gets at the sap by drilling a series of holes around the trunk of the tree. The holes are about 4 cm (1½ in) from one another and the rings of holes perhaps 10 cm (4 in) apart. It drinks the oozing sap by letting it run into the lower part of its beak. A wide variety of live trees are attacked. One elm was discovered with over 400 sap holes in it; and pines appear to be particularly favoured. If the tree is large the birds may concentrate their attentions on the sunny side where the sap rises most rapidly.

The megapodes of Australia use their tongues to monitor and regulate temperature. These birds build huge nests of compost inside which their eggs are incubated by the heat generated from the rotting plant material. The male cares for the nest and controls its temperature by probing with his beak and tongue, adding or removing vegetation as necessary.

*This young imperial eagle gives us a rare opportunity to examine a bird's tongue.*

Throughout much of its range, where food is reasonably plentiful all the year round, the great spotted woodpecker is resident and does not migrate. But in the more northerly regions such as Scandinavia, where it feeds extensively on the seeds of pines, it may behave in a different way. If the seed crop is poor in a particular area, vast numbers of woodpeckers may move, en masse, in search of new food supplies. Most of these temporary migrators are young, first-year birds, suggesting that they may be forced out by the older, more dominant individuals when food is scarce.

Since pine seeds are protected in a cone, the woodpecker resorts to specialised techniques in order to extract them. The cones are plucked or ripped off the tree and then carried to an 'anvil'. This is usually a crevice, deliberately chiselled out in the bark or branches of the tree, in which the cone is wedged for removal of the seeds. Alternatively the bird may use a natural cavity or even just a patch of hard ground. In the *Survival* film of great spotted woodpeckers one particular telegraph pole had evidently been used as a makeshift anvil for many years, and below there was a pile of discarded cones a metre high. An individual woodpecker's territory may contain a large number of anvils: one bird we filmed had 57 on a single tree, and another 32, of which only four or five were regularly used.

On collecting a pine cone, the woodpecker flies to a suitable anvil. If there is already a cone in the anvil it will hold the new one between its chest and the tree while it removes the old one. It then places the new cone upright in the crevice and proceeds to hack off the scales, extracting the seeds with its tongue. If a cone does not fit into the anvil, the woodpecker will chisel out more slivers of wood until there is enough space.

The time needed to remove the seeds depends on the type of cone. It takes the woodpecker about five minutes to extract all the seeds from a Scots pine cone. The Norway spruce, however, is a different proposition; the bird may spend up to half an hour on a single cone and still be able to remove less than half the seeds. Coniferous trees have, in fact, evolved different forms of cone partly as protection against the specialised feeding activities of birds like woodpeckers and other predators such as squirrels.

The anvil may also be used to deal with other food items, including hazel nuts, plum stones and marble galls. The last, sometimes known as 'oak apples', are formed when a tiny wasp lays its eggs on an oak twig. In response the tree produces an odd growth to contain the egg. Eventually the gall contains a grub, which makes a tasty meal. The woodpecker may even dismember a young bird at the anvil prior to feeding it to its own chicks.

Great spotted woodpeckers lead solitary lives during the winter, males and females settling in separate areas which vary in size, depending largely on the nature of the habitat. Where there is good feeding, as in mature deciduous woodland, a bird may occupy an area of a few hectares, but if the food potential is poor it may range up to a few kilometres from roost sites to favoured feeding areas. Often the same birds use the areas repeatedly, for as long as eight successive years.

——————————— □ ———————————

With the arrival of the breeding season (usually from January onward in the northern hemisphere) the birds start to pair up. The sexes attract each other by means of drumming and treetop calling. The rapid bursts of drumming, like those of a machine gun, are unmistakable. Yet they differ in pitch and rhythm from the tappings of other woodpeckers, providing immediate species identification. Furthermore, within the species, each bird drums in a slightly different manner, so that great spotted woodpeckers in adjacent territories can probably recognise their neighbours in this way.

The very survival factors which evolved initially for the purpose of obtaining food – powerful beak, reinforced skull and climbing prowess – are similarly employed here for locating and attracting a mate. Both sexes drum together as part of their courtship ritual, usually selecting a dead branch (or even a telegraph pole) which provides plenty of resonance, producing a sound which can be heard for a distance of up to 1 km (⅝ mile). The drumming sound is unmistakable and quite different from the hammering noises associated with feeding and nest-hole excavation.

*A juvenile male great spotted wood-pecker seen here at its anvil. An anvil is a kind of work surface used, for example, to extract seeds from pine cones.*

The beak moves so rapidly that the head appears as a blur. Indeed, high-speed film of the great spotted woodpecker shows that the bill strikes the hard surface 18–20 times each second. Each separate burst of drumming lasts less than a second, but some birds, especially at the peak of courtship activity, may drum ten times a minute. More curious is the occasional habit of 'silent' drumming, in which the woodpeckers go through the motions but do not actually touch the wood with their beak. The function of this behaviour is not known.

The main reason why the woodpeckers do not concuss themselves as a result of this continuous rapid drumming is that the skull is specially reinforced. There is a cushion of soft cartilage between the upper part of the bill and the skull, which acts as a natural shock absorber; and there is little space between the brain and the skull-case so that the brain does not bounce around and get damaged.

The pair also indulge in drumming when selecting a site for the nest. As a rule, it is the male who drums on one of his habitual posts, thereby

*The woodpecker does not get a splitting headache when it hammers wood for two reasons. The brain is in a muscular sling inside the skull. The upper beak is hinged, with strong muscles and shock absorbers to take the blows.*

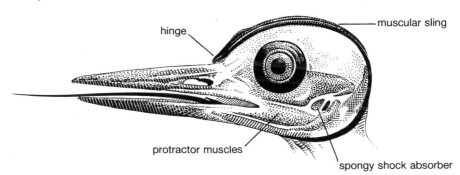

hinge · muscular sling · protractor muscles · spongy shock absorber

attracting the female. As the pair fly off towards the prospective site, the male performs a distinctive fluttering flight. If they find an old, empty nest they simply move in, but they may have no choice but to construct a new one. In the latter case, the birds first do some exploratory tapping around the opening. If the site proves suitable, they start to excavate, modifying the entrance until it measures 5–6 cm (2–2¼ in) across. Then both birds go to work on the nest proper, hacking out a chamber which is about 30 cm (12 in) deep and approximately 12 cm (5 in) in diameter, just enough for the birds to turn around. The male bears the brunt of the excavation work, chipping away for up to five hours at a time. The nest may take two to three weeks to complete, depending on the hardness of the wood.

During the nest-building activities the pair copulate frequently, perhaps up to six times a day. The female solicits the male with three short bouts of drumming and both birds utter loud calls during the actual mounting, which lasts only a few seconds.

Great spotted woodpeckers are monogamous, the male helping the female to rear the young. In Britain and central Europe egg laying occurs in late April or May. There is just one brood each year, but if the eggs are lost to predators, the pair will lay a replacement set. The clutch, laid on a layer

*A rare picture of the inside of a great spotted woodpecker's nest. Enough suitable trees for nesting are always critical for woodpecker populations.*

of chippings, usually comprises five or six white, slightly shiny eggs, each weighing about 5 g (¼ oz). As is the case with many birds, incubation commences only when the last egg has been laid, thus ensuring that all the chicks hatch simultaneously. Incubation lasts about twelve days, the male spending rather more time on the eggs. Each bird takes turns to do a shift that lasts 40–50 minutes, but the male, as in other woodpeckers, always takes the night shift. The young hatch blind and naked and crouch on the bottom of the hole, waiting to be fed by both parents on caterpillars and other insects. They leave the nest after three weeks, dispersing to lead independent lives.

179

# COOPERATIVE BREEDING IN BIRDS

The cooperative breeding behaviour of the acorn woodpecker is not exceptional. Nor is its habit of recruiting non-breeders to help in the rearing of the young. Some 200 bird species, mainly tropical, breed in groups; there are 60 in Australia alone.

Systems of cooperative breeding vary. For example, the superb blue wren of Australia breeds cooperatively in groups which consist of one breeding pair and several helpers, usually related males from an earlier breeding season. And the Australian magpie breeds and lives in groups which defend a common territory, but the additional birds rarely provide much assistance in tending the young.

Biologists had long been puzzled as to why certain birds should sacrifice their own breeding opportunities and act only as helpers. There seemed no good reason why these birds should not go off and produce their own young. They then discovered that in some species there were only limited opportunities for young birds. In some species, like the Florida scrub jay, the limiting factor is space. In others, like the superb blue wren, the limiting factor is females. This species breeds in solitary pairs when females are abundant. As a rule, however, since females suffer a higher death rate than males, there are rarely enough females to go round. Young males unable to find a female simply stay at home and help their parents. This has its benefits as a greater number of offspring are produced. Moreover, since helpers are closely related to the birds they help, and some of their genes are in the offspring, this 'helping' is a better option than not breeding at all.

*In breeding areas where food is hard to find, the male African pied kingfisher will recruit surplus males to help feed the chicks. Where food is plentiful no helpers are required.*

Although the great spotted woodpecker and most other woodpeckers are not noted for their social tendencies, leading solitary lives, one interesting exception is the acorn woodpecker. This species, which lives in California and ranges southward through Mexico and Central America to Colombia, exhibits one of the most intricate social relationships yet studied, living and breeding in groups. This woodpecker, as its name suggests, also has the unique habit of harvesting vast numbers of acorns and storing them in

*The special acorn storage holes of the acorn woodpecker are referred to as 'granaries'. Here a group of birds has converted a fence post into a store.*

specially constructed granaries. At the end of the 15th century a Spanish explorer in Mexico wrote of 'some thrushes called carpenter birds' which dug holes in large pine trees, placing an acorn in each hole 'so tight that they cannot be removed by hand'. But the truth was to prove even stranger. For the granaries, made up of thousands of individual holes, may accommodate some tens of thousands of acorns stored to be eaten throughout the year, particularly during the winter months.

181

*The acorn wood-pecker's granary is used to store many thousands of acorns for winter supplies.*

The best studied acorn woodpeckers live in the warm, dry woodlands of California, where there are five different species of oak tree, two evergreen and three deciduous. The birds normally form groups of about ten individuals, occupying and defending large territories, each of which contains one or more granaries. They store acorns in the autumn, from September to December, when they are plentiful, and feed on them through the winter until the supplies run out. In the breeding season, April to July, their diet consists largely of insects.

A concerted group effort is needed to construct a granary, fill it with acorns and protect it from marauders. An average granary may contain several thousand holes, each requiring at least 45 minutes to excavate. This may involve the group in many months of work. In mid-winter the granary will be crammed to bursting, with up to 3,000 acorns. There are reports, however, of certain groups which possess enormous granaries accommodating over 50,000 acorns. The capacity of the granary and the number of acorns per woodpecker have a marked effect on the size of the group and its breeding success. More acorns mean more group members and greater breeding success.

The typical group is made up of breeders and non-breeders. The breeders are usually closely related. Males may be brothers, or fathers and sons, while females are sisters or, more rarely, mothers and daughters. But there is seldom a close relationship between the sexes, which is important because in-breeding, as in any species, could produce harmful genetic defects. The females lay their eggs in a communal nest and there is close cooperation among them during incubation and subsequent rearing of chicks. Non-breeders are generally offspring from previous years and they too help to tend the newly hatched young. Later they may wander off to breed in other groups, or alternatively inherit the right to breed in their home territory, as might happen if all group members of the opposite sex were to die. Only on the death of its mother would the young male be free to breed with a new female.

*Some woodpeckers have adapted to treeless areas, such as this North American flicker, which has made its nest in a large cactus in the Sonoran desert of Arizona.*

Although the vast majority of the world's woodpeckers depend on trees for their survival, a few are specially adapted to treeless areas, and have therefore adopted uncustomary habits. The Andean flicker, for example, is found in mountainous regions above the tree line, feeding on the ground and nesting in rock crevices. Another, the carpintero, lives on the open pampas of South America. Charles Darwin described this species but his claim that the woodpecker never visited trees was challenged by W. H. Hudson, later a founder member of the Royal Society for the Protection of Birds, who pointed out that this species was capable of climbing and in some areas was especially dependent on the solitary ombu tree.

It is still debatable whether the pecking and drilling activities of woodpeckers damage trees. In forests where they have no alternative but to exploit sound, healthy wood, there may be some scarring of tree trunks; in general, however, they choose wood that is dry or already damaged by fungi and insects. There is no evidence, therefore, as once was alleged, that they represent a serious threat to the woodland environment.

*This tree bears the marks of excavation by a black wood-pecker.*

# Eagles

High in the Swiss Alps, the mountain slopes glisten in the spring sunlight. The warmth has lured a colony of marmots from their burrows. They are feeding in patches of vegetation among the rocks and engaging in playful fights with one another. Every now and then one animal stands upright on its hindquarters, darting glances in every direction, alert for any sign of danger. Suddenly it lets out a shrill whistle of alarm. Its companions immediately stiffen, remaining motionless and watchful. For a few moments all is still and silent. Then the great shadow of a golden eagle moves across the snowfields. The sentinel utters an urgent warning, and there is a mass of scurrying movement towards the burrow entrances. Within seconds the marmots are safely underground – all save one, which has lagged slightly behind. The eagle has spotted it. Closing its wings, the predator stoops at high speed, flattens out when only inches from the ground and grabs the marmot in its talons. But the little rodent manages to break free and tries to make a dash for freedom down the mountainside. The eagle gives chase, overtakes the marmot and strikes again. There is a brief struggle as hunter and prey tumble down the slopes together. This time the huge bird tightens its grip and the marmot goes limp. The eagle flaps its leisurely way to a nearby pine tree to devour its meal.

Eagles are birds of prey – a group that includes the hawks, falcons and New World vultures. Eagles are part of the hawk family, which has 217 members. Renowned for their graceful and strong flight, eagles can be very large: the harpy eagle weighs over 7 kg (15½ lb) – as heavy as a swan.

*Large birds of prey, such as this European buzzard, have distinctive wings with characteristic 'fingers' at the end of them.*

*Opposite The martial eagle is one of the largest eagles in Africa preying on monkeys, hyrax and small antelopes.*

Since ancient times eagles have been regarded as symbols of strength, courage and power. They were depicted on Roman coins and medals; and the bald eagle is the emblem of the United States. Today some 60 species occur all over the world, with the exception of Antarctica. In North America the bald eagle and the golden eagle are the most familiar species, while in northern Europe the golden eagle and the white-tailed sea eagle are the most common. The greatest number of different species is to be found in the tropics: Africa alone claims more than 20, including the huge martial and the unusual bateleur eagle with its very short tail. In Australia the most widespread species is the wedge-tailed eagle.

Many birds of prey are adapted for soaring flight, with broad wings to give extra lift. The larger eagles have huge wingspans, up to 2.5 m (5½ ft), which end in characteristic 'fingers'. This arrangement of feathers at the tips of the wings prevents stalling, especially when the bird comes in to land. For such large birds, flapping flight is expensive in terms of energy, particularly when covering great distances in search of food.

*As the sun heats the earth, the hot air rises from it, creating thermals. Birds of prey soar in thermals to gain height. They can travel long distances without effort by gliding from one thermal to another.*

The size and shape of the wings and, above all, the low 'wing loading', that is the large wing area relative to the body weight, allow the birds to soar without apparent effort for hours. They are assisted in this by columns of warm air – thermals – that ascend from the ground as it heats up in the sun. Having found suitable rising currents of warm air, eagles set their wings and fly in circles, rising higher and higher. Then they begin a gradual descent, during which they search for prey. Eventually they find a new thermal and the process starts again.

Thermals occur most frequently in the tropics but they form everywhere over large land masses in the summer. There are no thermals over the sea, which does not heat up as rapidly as land, and so when eagles migrate between Europe and Africa they do their best to avoid it. The birds cross the Mediterranean where it is narrowest – at the Straits of Gibraltar, and the Bosphorus at Istanbul. At both places, the skies are full of eagles at the time of the spring and autumn passage.

———————————— □ ————————————

Among the majority of eagles, as in most birds of prey, the female is bigger than the male, and thus the dominant partner. Pairs are monogamous, staying together year after year; and with the exception of some sea eagles, which occasionally breed in loose colonies, most species are solitary, each pair nesting on its own.

The density of breeding eagles varies according to species and region. In Africa, one pair of lesser spotted eagles may occupy an area of 10 sq. km (3.8 sq. miles), whereas pairs of martial eagles each need a space of more than 200 sq. km (77 sq. miles), with adjacent nests some 40 km (25 miles) apart. In general, the larger the species, the greater the territory required. Thus the black eagle, which weighs about 4 kg (8¾ lb), has a home range similar in extent to that of the martial eagle, while Wahlberg's eagle, weighing only 1 kg (2¼ lb), can survive in a much more restricted range, about the same as that of the lesser spotted eagle. Density is related to the feeding requirements of each species, and the number of eagles in any region during the breeding season depends on the food supply.

Outside the breeding period, too, eagles will congregate where there is abundant food. Every year several thousand bald eagles gather along one river in Alaska that remains unfrozen all winter and provides them with a plentiful food supply in the form of spent salmon – those which have spawned upstream and drifted down in a state of exhaustion. Many die, and the easy pickings attract the hordes of eagles.

Eagles site their large nests, or eyries, on cliffs, in trees and occasionally on man-made structures such as electricity pylons. Rarely, a nest may be seen on the ground. As a rule the sites chosen offer protection from predatory mammals and reptiles.

Many eagles build enormous stick eyries. In Scotland, some golden eagle nests may measure 2 m (6½ ft) deep and 3 m (10 ft) across. In most species it is the female who does most of the construction work, either collecting dead sticks or wrenching branches off trees. It is difficult to understand why eagles build such massive nests. Possibly they are more conspicuous – and thus a warning to other eagles that the territory is occupied.

*Bald eagles tend to return to the same nest sites. This nest is large and bulky and has probably been added to over many years.*

189

A single pair of eagles may construct several nests over the years. Among Scottish golden eagles, five nests per territory seems to be the average, although one pair was observed to have no less than fourteen. At the beginning of each season a few of these nests may be refurbished before the pair decides in which one to breed. The advantage of building several nests is that if the birds are disturbed they can quickly switch from one to another. If they had to start from scratch, following an interruption, they probably would not be able to breed that season.

Several species of eagles bring in sprays of green leaves as a finishing touch to their eyrie. Foliage is added before the eggs are laid, and may continue to be introduced throughout the breeding period, even when there are chicks in the nest. In the case of the African martial eagle, the nest cup is completely lined with green sprays prior to laying, whereas in others, like the harpy eagle, the entire nest may be covered in leafy twigs. The purpose of this behaviour is obscure, but it may also be part of the nest advertisement system, indicating to other eagles that it is occupied. There is another possibility. The green leaves may emit insecticidal substances, thereby keeping parasites under control. This has been proved in the case of the starlings which add greenery to their nests. The same could be true of eagles, but nobody has yet investigated this idea.

Territorial display reaches a peak in the weeks before egg-laying. The male usually performs conspicuous flights, chasing off rivals. If his threatening attitude fails to deter the intruder, fierce fights may ensue. The two birds will grapple and sometimes fall to the ground with their talons locked together. Fights to the death have been recorded among golden eagles.

Courtship activities include aerial displays by one or both partners. The pair may merely circle together over their territory, but sometimes they will perform elaborate and acrobatic manoeuvres. Male and female bateleur eagles perform spectacular twists and tumbles in the sky as a prelude to mating. The male crowned eagle soars up on a thermal, sometimes so high that he can hardly be seen with binoculars; then calling loudly, he descends rapidly in a series of spectacular stoops and climbs – this is known as the roller coaster display.

———————————— □ ————————————

The majority of eagles mate on or near the nest. As the male mounts the female, he is careful not to injure her with his sharp claws. Like most other raptors, they copulate frequently. A falcon – the European kestrel – holds the record, mating more than 600 times for each clutch of eggs; and the goshawk also copulates 500–600 times. These matings may occur over a period of two to three months prior to egg-laying, but even in the days just before laying, the birds can copulate five or six times a day. Data concerning eagles are less complete, but most species observed in detail mate at least twice a day for several weeks before the eggs are laid. This

# THE BATELEUR

The bateleur, which inhabits Africa south of the Sahara, resembles no other eagle, and is instantly recognisable by its spectacular black, white and chestnut plumage (not attained until its sixth year). In the air it is just as easily identified by its silhouette. The wings are usually long and the tail very short, so that in flight the red feet protrude.

The powerful wings, which contain 25 secondary flight feathers, more than in any other bird, enable the bateleur to soar on thermals to great altitudes. Yet it hardly appears to be using the wings at all for conventional flying. The name *bateleur* means 'tumbler' in French; a fitting name for in flight this eagle casts from side to side, using its long wings to help it balance in the air – like a tightrope walker using a balancing pole.

The bateleur may cover several hundred miles every day in search of food, which includes mammals and medium-sized birds such as hornbills, rollers and guinea fowl. In addition it eats reptiles such as tortoises, lizards and even venomous snakes. Sometimes, however, the biter is bitten; there is a report of one bateleur having been killed by a puff adder.

The speed and stamina required to travel such vast distances are equally evident in the eagle's nuptial displays during the breeding season when both sexes perform astonishing aerial manoeuvres. The male dives at the female from a great height, flapping his wings and calling loudly. As he approaches she flips upside down, presenting her feet to him, and then rights herself as he flashes past.

*The bateleur eagle is unmistakable in flight with its long upturned wings and short tail.*

*Most eagles actually kill their prey with their talons rather than with their beaks. When mating, the male 'balls' his feet to avoid injuring the female.*

level of sexual activity seems remarkable, given that a single mating is sufficient to fertilise all the eggs of the clutch. Moreover, certain other birds, such as the skylark, mate perhaps only once for each clutch. Why, then, do eagles mate so often?

One explanation is associated with the role adopted by each sex in the breeding season. Typically, the female stays on the nest while the male goes hunting on her behalf. He is therefore obliged to leave his partner on her own for long periods, during which time she could mate with another male. We do not know how frequently this happens but there have been several observations of strange males visiting unattended females and trying to mate with them on the nest. Repeated copulation by the male may therefore insure against his partner being unfaithful. It is not in his interests to work all season to feed his mate and her chicks if he is not their true father.

Another explanation is related to the pair bond. Matings that take place a few months before egg-laying are unlikely to fertilise eggs, so most probably they play a part in reinforcing the link between male and female.

Eagles generally lay a clutch of two, or rarely three, eggs. Although they may appear enormous, the eggs represent only a small proportion of the female's body weight. The white-tailed sea eagle, for example, weighs around 6 kg (13¼ lb) and each of her eggs is about 140 g (5 oz), a mere 3 per cent of her weight. Compare this with a smaller bird of prey like the tiny red-footed falcon which weighs only 150 g (5¼ oz); each of her three or four eggs is 17 g (just over ½ oz), or 12 per cent of her weight. (For more information on the importance of egg size see the box on page 139.)

Eggs are laid two or three days apart and are incubated for as long as six weeks. Usually this is the task of the female but in the tawny eagle the male may participate, and in the black eagle both sexes share incubation during the day, although the female alone does the night shift. The male golden

*This bald eagle chick will be tended by its mother in the nest for six to ten weeks.*

193

## CAINISM AMONG EAGLES

An eagle clutch generally consists of two eggs, yet in some species only one chick fledges. This is because the first-hatched young, 'Cain', kills the other either by depriving it of food and starving it, or by pecking it to death. Cainism, or siblicide, is not confined to eagles and happens in a few other groups of birds, notably certain seabird species. The killing of chicks by older siblings occurs routinely among black, lesser spotted, harpy, Wahlberg's, crowned and tawny eagles; and it has been observed, too, in golden and African fish eagles.

Why are such drastic measures necessary? The theory most frequently advanced is that it constitutes a form of brood reduction and is therefore a survival factor. Brood reduction is generally thought to occur when food is short, for it is better that one strong chick should be produced rather than two weaklings. Yet brood reduction, without resort to siblicide, happens naturally in many bird species. Among magpies, for example, the oldest or biggest chicks grab most of the food and the smaller ones die if there is not enough to go round. Among eagles, however, Cainism may occur even when food is abundant. The reasons why this form of behaviour occurs in eagles is as yet unclear. However, a recent study shows that siblicide is found primarily among those eagles where competition for breeding opportunities is unusually intense. In such cases it makes sense to produce a few high-quality offspring capable of passing on their genes to the next generation. By killing its nest mate a young eagle will get all the food its parents can provide, increasing its chances of growing into a healthy individual.

*An older eagle chick will sometimes kill its younger sibling by depriving it of food and starving it or by pecking it to death.*

Opposite *The African crowned hawk eagle is able to kill prey as large as antelopes. Here it has killed a young Thomson's gazelle.*

eagle does not even provide his incubating partner with food; instead, she leaves the nest for short periods in order to feed, possibly eating carrion rather than spending valuable time hunting live prey.

The chicks are tended in the nest for six to ten weeks, usually by the female, who shields them from the sun and rain, while the male hunts for food and brings it back to the nest for her to serve to the young. The eagle's powerful bill, which rips the food apart, can also be used delicately to offer tiny morsels to newly hatched chicks.

*Birds of prey, such as these golden eagles, often exhibit a size difference between the sexes, or sexual dimorphism. In most raptors, the female is larger than the male.*

♀        ♂

Parental duties do not necessarily end once the chick has left the nest. Most eagles look after their young for some time after they are fledged, and as a result raise only one brood each year. Some species have even longer breeding cycles, rearing chicks once every two years. The enormous crowned eagle lays one or two eggs, which are incubated for seven weeks. The young then remain in the nest for a further 16 weeks, and this is followed by an 11-month period of post-fledgling care. The entire cycle lasts one and a half years. A similarly protracted breeding cycle occurs among several other large species, such as the Philippines monkey-eating eagle and the South American harpy eagle.

*A Verreaux's eagle snatches up a rock hyrax.*

*This magnificent bald eagle exhibits two of the most important survival factors, for birds of prey: a strong curved bill and forward-pointing eyes, which give it binocular vision.*

The range of food taken by eagles as a group varies widely, but each species specialises. The diet of the biggest of all eagles, the Philippines monkey-eating eagle, is confined to monkeys and flying lemurs. Sea eagles snatch fish from the water. Golden eagles hunt mammals such as hares and rabbits as well as birds like grouse. Short-toed eagles feed primarily on snakes. Bateleurs and tawny eagles often operate as pirates and attack other birds of prey, forcing them to drop their catch. One species, the African vulturine eagle, is virtually vegetarian, subsisting almost entirely on the nuts of the oil palm, supplemented by the odd crab or dead fish. Another with an unconventional diet is the Indian black eagle which preys primarily on the eggs and nestlings of other birds. Although little is known of this

## SNAKE-EATING EAGLES

Snake eagles, serpent eagles, harrier eagles, the bateleur and the white-bellied sea eagle are all adapted for catching and eating snakes. Most of these species have short, rough toes with a powerful grip – essential for grappling with their prey. The majority of snake-eating eagles inhabit open country, but the serpent eagles, with unusually large eyes, are attuned to hunt in the dim light of tropical forests. They are sit-and-wait predators, watching for the give-away movements of snakes or chameleons. The crowned serpent eagle specialises in tree snakes, and the white-bellied sea eagle of south-east Asia feeds on highly venomous sea snakes.

Despite their food preferences, it is unlikely that these eagles are immune to snake venom. But they are often well protected. The short-toed eagle, a summer visitor to Europe and particularly common in France, Italy and Spain, has exceptionally thick scales on its legs and dense down on the underside of its wings. When catching a snake, it spreads its wings to distract the reptile from its body, for a bite on the wing will do little harm. If the snake is small, the eagle seizes it in its talons, carries it wriggling into the air and transfers it to the beak, which crushes the reptile's head. Bigger snakes – up to 1 m (3 –4 ft) in Europe and 1.8m (6 ft) in India – are tackled on the ground. Some are broken up and consumed piecemeal, others are swallowed inch by inch. Undigested scraps of skin and scales are coughed up as pellets.

The short-toed eagle even juggles with a snake during his courtship ceremonies, carrying it in his bill as he soars over his territory and offering it to the female in flight.

The young eagles are fed on snakes, regurgitated from the parent's crop. On one occasion in the Coto Doñana, in southern Spain, a male brought up a snake 1 m long for its chick. The female started to tear it to pieces but the chick, less than half the reptile's length, seized the snake and began to swallow it head first. At one point the weight caused the chick to topple over, but it soldiered on. Three hours later the chick was ready for another.

*The short-toed eagle specialises in eating snakes: its legs are covered with thick scales and the underside of its wings with dense down as protection against venom.*

species, it has been suggested that the foot and claw formation, with claws only slightly curved, allow a wider grasp, so enabling the eagle to grab complete nests as it flies past.

Female eagles tend to be larger than male eagles, and often this size difference can be marked. One consequence of this is that male and female often take prey of different sizes. This may help to prevent the two sexes competing for food in the same territory.

The largest eagles require approximately 3–6 per cent of their own body weight – about 200 g (almost ½ lb) – of food every day. The amount depends on how active they are, and the ambient temperature. At low outside temperatures, as with human beings, food requirements tend to be greater, since more is used in maintaining body temperature.

Like all raptors, eagles possess acute vision. Their comparatively large eyes enable them to spot prey from considerable heights. Having sighted it, they dive with unerring precision. The animal is grasped in the immensely strong talons, which can contract and relax rhythmically at high speed, puncturing the victim's vital organs and eventually killing it. The largest eagles can successfully tackle animals weighing up to about 7 kg (15½ lb), although one martial eagle is reported to have killed a fully grown suni, a small African antelope weighing over 30 kg (66 lb).

After killing their prey, eagles normally remove the skin, the stomach and the entrails, and feed on the muscle – the meat. Unless they are really hungry, they tend to be fussy, eating only the best parts. When hunting for the chicks, eagles carry food back to the nest. The amount that they can hold has been the subject of much speculation and fantasy. A golden eagle was once seen with an adult fox in its talons, while a harpy eagle was observed clutching a 7 kg (15½ lb) sloth. There is a also a report of an American bald eagle lifting a 4 kg (8¾ lb) greylag goose. Stories of eagles snatching up children, popular in folklore, are generally fictitious.

———————————— ▢ ————————————

Eagles usually produce only one chick each year, which in turn does not normally breed until it is between four and nine years old. But the birds may raise a number of offspring because they live a long time. The record ages for various species in zoos are: golden eagle, 48 years; white-tailed eagle, 42; Steller's sea eagle, 32. The bald eagle may live for up to 36 years in captivity, but in the wild its average life expectancy is 20 years. The oldest eagle of all, however, is reported to be a bateleur aged 55 and still going strong.

Natural causes of death among eagles include starvation, disease and predation. But human persecution clearly increases the toll. An investigation of 231 bald eagle deaths in North America between 1966 and 1974 showed that almost half had been shot for 'sport', 20 per cent died of accidents, 10 per cent from pesticide poisoning, 12 per cent from natural causes and 14 per cent for unspecified reasons.

*A golden eagle on a red deer calf carcass. Eagles are attracted to carrion.*

Opposite *This tawny eagle is just closing its third eyelid or nictitating membrane, which is thought to give it extra protection while attacking its prey.*

Birds of prey, but particularly eagles, have suffered for centuries because of man. Many species have been condemned without fair trial for hunting domestic lambs. In fact, they scavenge the carcasses of dead lambs but only once in a while take a live one. But sheep farmers have gone all out for the golden eagle in Britain and North America, for the white-tailed sea eagle in Greenland and Norway, for the wedge-tailed eagle in Australia, and for the martial and black eagles in South Africa.

Eagles were – and still are – killed in a variety of ways, shooting, trapping and poisoning being the most common. In many countries, steel-jawed gin traps would typically be placed atop specially built mounds of stone. The eagle, caught by the feet, usually suffered a terrible lingering death. These

traps are now generally illegal, but this does not prevent their use. Before it was banned, the most commonly used poison in Britain was strychnine, whereas in North America and Australia a poison called 'Ten-eighty' (sodium monofluoracetate) was more widely used. This was smeared on animal carcasses, mainly to kill coyotes, wolves and the like, but inevitably poisoned eagles, such as the wedge-tailed and bald eagles, as well. In Texas, golden eagles were shot from aeroplanes, a method which killed several thousand birds over a period of 20 years before it was banned in 1962.

*Many birds of prey, like this sparrow-hawk, have died at the hands of man. Pesticides and persecution are still major problems. Although many pesticides are not used in Britain and America, they are still manufactured and exported to other countries in Africa and elsewhere.*

*Pallas's fish eagle.*

Because the killing of eagles, and other birds of prey, was often associated with 'bounty' payments, there are depressing reports of heavy slaughter in particular areas. In Western Australia, bounties were paid on no less than 147,237 wedge-tailed eagles between 1928 and 1968. Even though the system was then discontinued, it is believed that 30,000 'wedgies' are still killed annually in Australia. This is especially ironic because the eagle is one of the few species capable of controlling the plague of introduced European rabbits there.

In Norway, bounties were paid on golden and white-tailed sea eagles, and in the latter half of the 19th century more than 61,000 of these two

203

*Opposite Birds such as this immature African fish eagle are very much under threat from pesticides. High levels are known to exist in these eagles on Lake Kariba in Zimbabwe.*

species were massacred. Bounties were still being offered there as recently as 1963. In Scotland, on one estate, the following birds of prey were killed in the three-year period 1937–40: 98 peregrine falcons, 78 merlins, 462 kestrels, 285 buzzards, 15 golden eagles, 27 white-tailed sea eagles, 18 ospreys, 63 goshawks, 275 kites and 68 harriers. The destruction was as devastating as it was widespread.

Once this deliberate killing had declined due to a change in public opinion, eagles and other raptors were confronted by a new source of man-made mortality: pesticides. The organo-chlorine based pesticides, such as DDT, have had a disastrous effect. These compounds are highly toxic, but also very stable, so that they remain unchanged in the environment for long periods. They also tend to accumulate in the bodies of animals, particularly in their fat, thus becoming concentrated in birds of prey at the top of the food chain. Even very low levels in a bird's body tissues can have drastic effects on reproduction.

DDT was first used to combat agricultural pests in Britain and North America after the Second World War. It was not until the 1960s, when the damaging side-effects became apparent, that its use was restricted. By that time a number of bird of prey populations had plummeted. The best-documented effects are of egg-shell thinning, which was especially prevalent among bird-eating and fish-eating raptors. As a result of pesticides, the thickness of egg shells of bald eagles, white-tailed sea eagles, peregrine falcons and ospreys was reduced in all cases by over 15 per cent. The consequence was that the eggs were crushed or broken as the birds tried to incubate them. Breeding success fell catastrophically. For example, in Sweden, of 130 peregrine territories occupied in 1940, only 10 were still being used in 1975 – a 92 per cent reduction.

In North America, the use of DDT peaked in 1959, but had virtually stopped by 1973. Just two years later the decline in bald eagle populations had been checked. In Britain golden eagles were affected by the pesticide dieldrin, which was used in sheep dips. Fortunately, dieldrin was banned in 1969. Although pesticides like DDT are rarely used now in Britain or North America, they are still manufactured in Europe and America but sold to, and widely used by, developing nations, including those of South America, North Africa, the Middle East, India and south-east Aisa. On a global scale, therefore, the total amount of organo-chlorine pesticides applied is not decreasing. Inevitably this means that birds of prey in those areas will decline in numbers.

In Britain, however, the eagles have had a slight reprieve. Since the ban on dieldrin, the body levels of this pesticide in golden eagles has dropped by half, and breeding success has doubled. In addition sea eagles, which became extinct in 1916, have been reintroduced with encouraging results. Although man has given them a helping hand, it now remains to be seen whether their habitat will support them. But one thing is sure: they will need every one of their special survival factors if they are to thrive.

# Index